Remediation and Carbon Capture in Mining Sector

José Marrugo-Negrete · M. P. Jonathan ·
AL. Ramanathan · S. Chidambaram · R. Nagarajan ·
Pankaj Kumar
Editors

Remediation and Carbon Capture in Mining Sector

Springer

Editors
José Marrugo-Negrete
Grupo de Agua, Química Aplicada y Ambiental
Departamento de Química
Facultad de Ciencias Básicas
Universidad de Córdoba
Córdoba, Colombia

M. P. Jonathan
Centro Interdisciplinario de Investigaciones y Estudios sobre Medio Ambiente y Desarrollo (CIIEMAD)
Instituto Politécnico Nacional (IPN)
Ciudad de México (CDMX), La Laguna
Ticoman, Mexico

AL. Ramanathan
School of Environmental Sciences
Jawaharlal Nehru University
New Delhi, Delhi, India

S. Chidambaram
Water Research Center
Kuwait Institute for Scientific Research
Safat, Kuwait

R. Nagarajan
Department of Applied Geology
Faculty of Engineering and Science
Curtin University
Miri, Sarawak, Malaysia

Pankaj Kumar
Water and Adaptation
Institute for Global Environmental Strategies
Kanagawa, Japan

ISBN 978-3-031-87882-4 ISBN 978-3-031-87883-1 (eBook)
https://doi.org/10.1007/978-3-031-87883-1

Jointly published with Capital Publishing Company
The print edition is not for sale in Afghanistan, Bangladesh, Bhutan, India, Sri Lanka, the Maldives, Nepal and Pakistan. Customers from Afghanistan, Bangladesh, Bhutan, India, Sri Lanka, the Maldives, Nepal and Pakistan please order the print book from: Capital Publishing Company.
ISBN of the Co-Publisher's edition: 978-93-81891-87-2

This Springer imprint is published by the registered company Springer Nature Switzerland AG
The registered company address is: Gewerbestrasse 11, 6330 Cham, Switzerland

Preface

With rapid environmental changes and global warming reaching 1.5 degrees higher than the industrialised era, several planetary boundaries are reaching their tipping points. Irregulated mining activities are one of the key drivers triggering these negative consequences on Mother Earth. Keeping this in mind, recently in COP and different global negotiations, most of the countries pledged to work towards achieving a net-zero society by the next few decades. Moreover, carbon capture and carbon storage become imperative to reduce GHGs emissions and can contribute to the transition to a net-zero society. This book presents a holistic picture of mining practices and its impacts on socio-environmental impacts followed by different state-of-the-art management techniques for carbon capture and governance, a way forward in achieving a net-zero carbon society. In the first instance, it presents the link between mining, climate change, decarbonization, and its applications. GIS and modelling techniques were brought out by the case studies from Columbia, India, Malaysia, and Mexico. First theme focusses on steel, nickel, and limestone mining and its impacts on various aspects of socio-economic health in the ambient environment. The studies on the theme '*traditional and modern practices in mining*' from these regions focussed on: steel manufacturing and its evolution in the practices in the production process, Geological effects on the fragmentation of Quarry blasting, and optimization of Pillar dimensions of an underground mine. Further, the second theme on '*environmental pollution due to mining practices*' highlighted pollution in sediments and water, GIS application to assess the impact of open-pit mining; water quality and pollution index; impact of Nickel mining on freshwater fishes, their impacts in food chain; impact of coal mining, gold, and chromite mining on water resources and soil in the adjoining regions. The studies were representative of cases in Latin America, Colombia, and India. Second theme focusses on direct and indirect risks and challenges to environment and human health/health risk in Colombia and India due to mining activities. One study focussing on contaminant fate and transport through food chain with an example of nickel enrichment in fishes and its bioaccumulation in the humans with a case study from Colombia. Another research on the status and extent on the impact of stone mining on human health in the northern state of Himachal Pradesh from India indicates that groundwater has reached an alarming stage of contamination and necessary steps are to be taken to prevent further contamination in the region.

Last, it discusses an array of topics related to carbon capture, neutralisation, and reclamation relevant for contemporary policy issues at global landscape. With different case studies, it highlights the role and possibilities of using mining waste for carbon storage and sequestration and the key factors controlling this removal process are its mineralogy, particle sizes, and grain size. Furthermore, it discusses the importance of Shock Wave Technology (SWT) and its feasibility in grinding the mining waste, which improves the CO_2 removal rate, reduces the land requirement for mining waste disposal, reduced operation and maintenance costs; all required for scaling it up at the industrial level. Followed by that, one study reviews state-of-the-art publications on the techno-economic evaluation of chemical absorption most advanced technologies employed for carbon capture. Subsequently, they revealed that removal efficiency varies with major parameters such as plant location, coal price, emission source, CO_2 concentration in the flue gas, storage location, and type, transport facility variety, solvent type, capacity factor, etc. Another chapter presents the importance of various reforms in Mexico, which are needed for sustainable management of mining sectors. Whether it is about (a) presenting an applicable regulatory framework like reforms to mining laws and the General Law of Ecological Balance and Environmental Protection (LGEEPA), binding with international agreements and negotiations, and strict enforcement mechanisms, or (b) technology reforms which can achieve the synergies between decarbonisation and minimising energy losses; which are the key elements to design various mitigation strategies for abating pollution, climate related risk and other negative effects. Next chapter presents the importance of geological sites such as saline aquifers and depleted oil and gas reservoirs which are the most practical options for CO_2 sequestration due to their limited operational costs for various trapping mechanisms and providing the potential for Enhanced Oil Recovery (EOR) as well. Finally, this section provides an overview of CO_2 reduction strategies in the fossil fuel electricity supply chain, in addition to CO_2-capture technologies. Finally, the last chapter discusses the role of environmental policy and governance for managing various issues from mining sector with a case study from Mexico. In summary, various practical and sustainable initiatives range from heavy investment in institutional reforms, stakeholders' engagements, participatory approach, etc. to accommodate all the necessary elements for sound implementation of decarbonisation policies.

Córdoba, Colombia — José Marrugo-Negrete
Ciudad de México (CDMX), Mexico — M. P. Jonathan
New Delhi, India — AL. Ramanathan
Safat, Kuwait — S. Chidambaram
Miri, Malaysia — R. Nagarajan
Kanagawa, Japan — Pankaj Kumar

Contents

About the Editors

Dr. José Marrugo-Negrete is the Director and Senior Professor at Instituto Regional del Agua (IRAGUA) and Applied Chemical and Environmental Water Research Group, as well as the Regional Director of the SUE Caribbean Master degree program in Environmental Sciences of Universidad de Córdoba, Colombia. He is a Senior Researcher recognised by the Ministry of Science and Technology of Colombia. His areas of interest include the dynamics of heavy metals in the environment, their impact on human health and environmental remediation. His research experience is based on studies of environmental contamination mainly by heavy metals and pesticides, and the risks to human health associated with exposure to them. He has also developed research on methods of elimination and reduction of environmental pollutants, with the aim of recovering contaminated ecosystems affected by mining. He has made significant contributions to environmental and mining policy development in Colombia.

Dr. M. P. Jonathan is a Professor at Centro Interdisciplinario de Investigaciones y Estudios sobre Medio Ambiente y Desarrollo (CIIEMAD), Instituto Politécnico Nacional (IPN) in Mexico City and is a member of National Researchers Scheme—Level 3. His research includes Environmental Geochemistry focussing on metal cycling in water, sediment, and air in continental/coastal environments and paleotsunami studies. A worldwide program called '*Blue Flag Beaches: Vision-2030*' dealing with the creation of a baseline data set on geochemical composition/enrichment of elements in tourist beaches is currently ongoing. Studies on mega and microplastic in beaches, coralline regions, commercial fishes, crocodiles, and sea lions have been initiated. He is serving in the Editorial Board of Environmental Pollution (Elsevier Publications), Environmental Monitoring and Assessment Journals (Springer-Verlag) and has published over 150 research articles.

Dr. AL. Ramanathan is a Professor in the School of Environmental Sciences (SES) at Jawaharlal Nehru University (JNU), New Delhi, India. He was in the top 2% in the field of Environmental Sciences in the world list from 2019. He has published over 250 research papers and his research includes river sciences, groundwater hydrology, mangrove ecosystem biogeochemistry, and Arctic, Antarctic and Himalayan research. He has supervised 34 Ph.Ds.; ~100 M.Phil. and Master Dissertation; ~20 postdoctoral fellows. He has served as Dean, SES at JNU. He was an arctic group leader for three years and participated in more than 30 national/international projects. He has been awarded the INSA teachers, Young Scientist Award—DST, Government of India; Young Scientist Award Lineaus-Palme Fellow, FORMAS Fellow—Sweden; SIDA Fellow—Sweden; CIFIPRA Fellow—France.

Dr. S. Chidambaram is a Senior Research Scientist at the Water Research Centre, Institute for Scientific Research, Kuwait. He is a gold medalist in undergraduate and postgraduate degrees. He served as Professor and Head from 2015 to 2017 with 19 years of research experience and graduated 18 postgraduate, and 18 doctorate students. He published around 250 research articles and 10 books. He served as guest editor for VSI in different journals and as Associate Editor in the Kuwait Journal of Science. Computer programs WATCLAST and CHIDAM were developed for Hydrogeochemical studies. He received the best paper award in the Applied Water Sciences journal in 2013 and Hydrogeology/Environmental Geochemistry in 2016 by the Indian Society of Applied Geochemists. He has undertaken training in IAEA programs and as an expert for various missions.

Dr. R. Nagarajan is a Professor in the Department of Applied Geology, Dean R&D and Director of Curtin Malaysia Research Institute. His research interests and capabilities include applications of geochemistry to various evolution of sedimentary systems, petroleum geology, environmental geosciences and biogeochemistry. He has published over 100 indexed publications (Journal papers, conference proceedings and book chapters). His current research focusses on the geochemistry of Oligocene to Neogene sedimentary/meta-sedimentary rocks and recent sediments of Borneo. He has been awarded the top most prestigious award called '*John Curtin Distinguished Professor*' from Curtin University Australia for significant and sustained contribution to the University a national/international reputation and a distinguished record of service to professional and public life. He has more than 10 years of leadership experience in the above administrative positions with Curtin University.

Dr. Pankaj Kumar is currently working as a research manager in the field of water resources and climate change adaptation at the Institute for Global Environmental Strategies (IGES), Japan. In addition, he is also working as head of IPBES-TSU on Scenarios and Models. Prior to this, he worked in United Nations University, Tokyo; for more than three years. His research work primarily focuses on Socio-hydrology, Water security, Hydrological simulation, Scenario analysis, and Water-health-food-energy nexus. He is also engaged in teaching, focussing on assessing the impact of global climate change on water resources and weighing on adaptation and mitigation options to offer scientific solutions. He has work experience with different global assessments like IPCC AR5, AR6; IPBES NEXUS assessment, GEO7, etc., as Chapter Scientist and lead authors respectively.

From Ore to Steel Manufacturing and Its Evolution in the Practices of Productive Processes in Mexico

María Concepción Martínez-Rodríguez,
Lorena Elizabeth Campos Villegas,
and Ana Laura Cervantes-Nájera

1 Introduction

Steel, an essential material in construction, is an alloy of iron (Fe) and carbon (C); sometimes, other elements, such as chromium (Cr) or nickel (Ni), shown in Fig. 1, are added for specific purposes to improve some of its properties. Research carried out by the steel industry has led to several groups of new steels that satisfy various properties, and there is now a wide variety of standards and specifications. For example, the Society of Automotive Engineers (SAE) classifies steel mainly into carbon, medium alloy, stainless, high strength, tool, and other steel (Black and Kohser 2008).

Steel contains a proportion of carbon ranging from 0.03 to 2%. The physical properties of steels and their behaviour at different temperatures depend mainly on the amount of carbon and its distribution. Since steel is highly refined iron, its manufacture begins with reducing iron, which is later converted into steel. Although pure iron is one of the elements of steel, it is not found free, as it chemically reacts quickly with oxygen present in the air to form iron oxide (rust). Rust is found in significant quantities in iron ore, a concentration of iron oxide with impurities and iron materials.

Steels can be classified by different systems depending on their chemical composition, manufacturing methods, applications, processing methods, product forms, heat treatments, and microstructure. The combination of alloying element effects and heat treatments produces a variety of properties and microstructures; its main phases are ferrite, austenite, and cementite (Robledo Monsivais 1997). Steels are divided according to their carbon content into the following categories:

- Eutectoid steels: this type of steel contains 0.76% of carbon.
- Hypo-eutectoid steels: this type of steel contains less than 0.76% of carbon.
- Hyper-eutectoid steels: this type of steel contains more than 0.76% of carbon.

To make steel, mineral raw materials, as shown in Table 1, are required, from which different chemical compositions are produced to generate the steel characteristics required to satisfy customer's demand (lighter, more substantial, or highly malleable steel). The elements used, carbon, boron, sulphur, chromium, phosphorus, molybdenum, cobalt, and magnesium, are obtained from these ores.

Few processes are more extraordinary than the generation of steel. It all starts with a tremendous explosion that destroys the earth's crust in magnitudes close to a soccer field to release or enter the crust's interior and reach each mineral deposit rich in the element required to manufacture the most demanded metallic alloy in the world (steel).

Mineral extraction is carried out in different stages: prospecting, extraction, beneficiation, and abandonment of the site and waste disposal (mining tailings). Specifically, extraction, irrespective of dry or wet methods used to carry it out, could be one of the mechanisms with the most significant environmental impact on the earth's surface; both alternatives whose environmental consequences have an effect are summarised in Table 2.

Owing to the large volume and complexity of operations and the high consumption of energy and raw materials, the iron and steel industry, similar to other 'heavy' industries, can have a significant impact on the environment and the population of neighbouring communities, air pollutants, wastewater pollutants, and solid waste. Steel production from iron ore is not a one-step process; according to Kalpakjian and Schmid (2010), it includes a series of stages that fall into the following

M. C. Martínez-Rodríguez (✉) · L. E. C. Villegas ·
A. L. Cervantes-Nájera
Instituto Politécnico Nacional (IPN), Centro Interdisciplinario de Investigaciones y Estudios Sobre Medio Ambiente y Desarrollo (CIIEMAD), Alcaldía Gustavo A. Madero, Mexico City, Mexico
e-mail: mcmartinezr@ipn.mx

J. Marrugo-Negrete et al. (eds.), *Remediation and Carbon Capture in Mining Sector*, https://doi.org/10.1007/978-3-031-87883-1_1

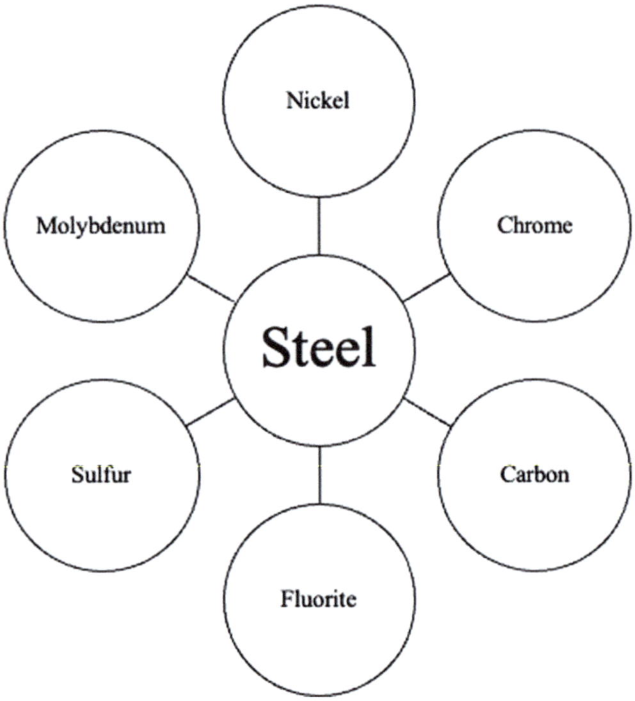

Fig. 1 Elements used for steel

Table 1 Raw materials and geological origin

Metallics	Geology
Chromite (Cr)	Magmatic segregation
Bornite and chalcocite (Cu)	Hydrothermal deposit, metamorphic, process enrichment
Hematite, magnetite and limonite (Fe)	Sedimentary formation
Pyrolucite (Mn)	Hydrothermal deposit
Pentlandite (Ni)	Magmatic segregation
Molybdenite (Mo)	Hydrothermal deposit
Scheelite and wolfranite (W)	Metamorphic deposit
Sphalerite (Zn)	Hydrothermal deposit
Apatite [$Ca_5(PO_4)$]	Sedimentary formation

(Arai 2021; Equihua et al. 2010; Han et al. 2014; Moe et al. 2018; Pandiand and Varma 2001; Wei et al. 2018; Zhang et al. 2021; Zhang et al. 2003)

three steps: (a) first step, melting, (b) second step, acceleration and (c) final step, finishing.

(a) First step, melting. As the present research proposes afterwards, its technological evolution is the one that historically has taken more time. As an introduction, we mention that it is carried out in a 'blast furnace'. However, the first furnaces reached 8 m in height; nowadays, they reach up to 35 m in height. Once turned on, the blast furnaces can last for even more than a decade in continuous operation; because of the complexity and cost involved in shutting them down for maintenance, they work 24 h a day, 365 days a year. Inside, iron ore in the form of iron oxide is mixed with coking coal; the coal reacts with the oxygen present, forming carbon dioxide and carbon monoxide, while the liquid iron ore, often called 'pig iron', accumulates at the bottom along with other residual minerals floating on the surface.

(b) Acceration. Sponge iron enters a basic oxygen furnace (BOF), often feeding an electric arc furnace (EAF), where a similar process is carried out. For steelmaking (steel generation), the liquid steel is moulded to obtain various semi-finished products, depending on its shape and end use, such as ingots, slabs, billets or plates. Materials are subjected to various manufacturing processes to obtain the final product, ranging from structural steel, rods, and other materials for construction, sheets or plates widely used by the household appliances and automotive industry, to products used in food production and surgical instruments.

Table 2 Environmental effects of mining

Environment	Dry extraction	Wet extraction
Land surface	The devastation of surfaces; alteration of morphology; the danger of landslides at the star-up face; and destruction of cultural property	The devastation of surfaces; alteration of morphology and river course; and formation of large waste dumps
Air	Noise and vibrations generally, noise and vibrations from detonations; dust formation from traffic and erosion; fumes (including fumes from self-igniting dumps); and blasting fumes, noxious gases	The noise produced by power generating equipment, extraction, treatment, and transport works; exhaust fumes
Surface water	Alteration of nutrient cycling (possible eutrophication); contamination with sewage; contamination caused by intensified erosion	Denitrification; contamination of the receiving watercourse with large quantities of study sewage and contaminated wastewater
Groundwater	Lowering of groundwater table; deterioration of groundwater quality	Alteration of water balance and groundwater quality

(Cortez-Lugo et al. 2018; Salcedo Sánchez et al. 2022; Schiavo et al. 2022; Wurl et al. 2018)

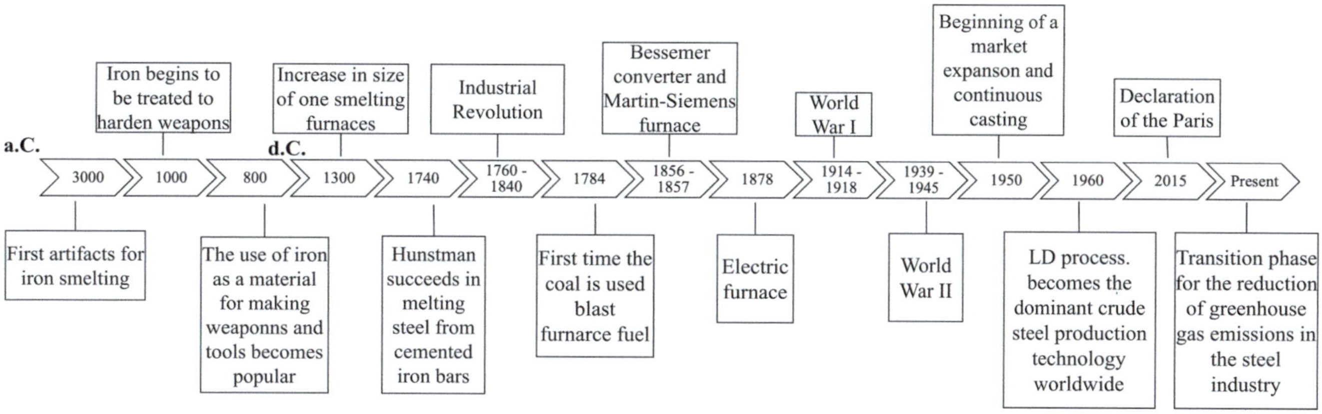

Fig. 2 Graphical representation of the chronological sequence of important events in the history of steel

(c) The final finishing of semi-finished products requires at least one of the following three processes: rolling, forging or casting. Rolling can be cold or hot, the most common being hot rolling. The semi-finished product is introduced for several times to a series of rolls to reduce thickness and generate rolled sheets (coils). Forging also modifies the semi-finished product to bring it to the desired shape through heating and pressure, usually using hammers to give it the form of a design. Finally, casting; as the name indicates, the semi-finished product is melted to cast into moulds of the desired shapes. Once the final products are obtained, an activity that on many occasions is carried out in the same plant that made the steel, they still have a long way to go until the final finishing that turns them into direct consumer products. These processes depend on the application to which the final product is subjected, such as galvanised, tinned, tempered, or annealed. It should be noted that steel plants are considered 'integrated' if they develop the above-described three basic processes for steel production.

This work aims to know the temporal relationship of the development of a natural resource, such as steel, concerning the technological evolutionary development of man, specifically in Mexico. For this purpose, a historical review of open online documents on government's pages and scientific articles was carried out as a methodology.

2 Historical Background of Steel

The date when the technique of smelting iron ore was discovered is unknown. The first artefacts found by archaeologists date back to 3,000 BC. in Egypt. However, the Greeks were hardening iron weapons by heat treatment around 1,000 BC. There is a historical period in which the use of iron for making weapons and tools was discovered and popularised: the Iron Age. The hardness and abundance of iron ore made it a much easier material to be obtained than bronze, which contributed decisively to its adoption as the most widely used metal (Moffit 1998).

From the fourteenth century onwards, iron ore was introduced at the top in large smelting furnaces and reduced to metallic iron, which absorbed more carbon as gases passed through it. Later, carburising was perfected by melting steel in clay crucibles, and from 1740 in England, crucible steels were obtained, a technique developed by Benjamin Huntsman. By 1856, Henry Bessemer invented a method for mass production of steel, but since only the iron-containing phosphorus and sulphur in small proportions could be used, it was abandoned, and all the forests in England had collapsed by then (Moffit 1998).

Owing to the Industrial Revolution in the seventeenth century and the rise of the railroad industry, the market's expansion, the steel industry's mechanisation, and constant technological innovations began. It would later emerge what was called mass production; these events are presented in Fig. 2 (Hobsbawm 2003).

3 Evolution in the Practices of Steel Production Processes in Mexico

In Mexico, since the sixteenth century, the traditional models of Spain were copied to work iron, with the so-called Catalan forge, which, with its double bellows, could raise the temperature much higher than the traditional forges and thus perform the material better, achieving more quality and hardness in the objects produced by the blacksmiths. This innovation transformed metallurgy, placing Spain in the sixteenth century among the most advanced nations in Europe in this field (Bolis-Morales 2017).

According to Soriano (2010), in 1808, the orchard of the San Gregorio College (San Ildefonso Street, Mexico

City), the seat of the Workers' University, was taken over by Manuel Tolsá (1757–1816) for the manufacture of 49 pieces of artillery (cannons) with their respective assembly parts. For this, Tolsá used the facilities already existing in the college; there were lathes, foundry ovens, boring machines, moulds, stables, and other parts that Tolsá built for his services. A multitude of workers worked in these workshops, such as Indians, journeymen, labourers, foundry workers, stonemasons, carpenters, moulders, lathe workers, filers, sharpeners, roofers, muleteers, watchmen, stablemen, and other types of works.

Along with the adaptation of Tolsá's workshops, blast furnace technology was established in Coalcomán, Michoacán, central-western Mexico, in 1807 by Andrés Manuel del Río (1764–1849). It began operating on 29 April 1807 with a large-scale reverberatory furnace that remained in operation until the independence movement. Thus, with the growing demand and capital investment, the companies replaced their old facilities, such as the Catalan furnace, with the most modern blast furnace technologies (Sánchez Díaz 2009). These were established and operated at places far from the large population centres but close to the iron deposits, where sufficient water and forest resources were present for the supply of materials and fuel to allow the exploitation and processing of iron.

After several trial-and-error experiences and overcoming difficulties, Andrés Manuel del Río realised that it was necessary to take into account the physical characteristics of the organic composition of the ferrous minerals used:

> From this fact, I inferred that there are metals that do not let melt in a hurry, for example, that of Coalcomán, which is a brown iron with much manganese, needs to melt slowly with iron, and perhaps gives it the endurance, consistency and hardness that characterise it and that there is strong iron with the same properties, which is not supposed to have any manganese. However, when it comes out grainy in the fracture and sour, as happens a few times, I think it is a real alloy of iron and manganese, perhaps with an excess of the latter. (Sánchez Díaz 2009)

This shows the evolution of steel production before the development of analytical chemistry.

In some cases, the forges that operated in the first half of the nineteenth century were simple foundries that processed ore on a low scale or produced low-quality irons used to manufacture agricultural tools to extract minerals and ironwork with beasts of burden. In 1847, Juan M. Flores promoted an essential technological improvement in the ironwork facilities by substituting the smelting system of Catalan forges for furnaces similar to those used at that time in the ironworks of Vizcaya. However, this technological innovation was not reflected in an increase in production. Despite this, the products and tools generated in the forge became more diverse and of better quality as they began to produce hoes, wheels for cars and carts and iron ploughshares, and parts for machinery and beaten iron objects (Sánchez Díaz 2009).

In the twentieth century, the forges were replaced by steel companies, and at the end of World War II, the growing demand of steel drove the emergence of several steel companies in Mexico in 1970, and with the creation of Siderúrgica de Las Truchas in Michoacán, western Mexico, the national steel industry was consolidated (Toledo Beltrán and Zapata 1999). In addition, at this time, the dominant process for producing steel was the blast furnace, but this was not the only process used to obtain the metal. In 1979, Juan Celada Salmón, a Mexican engineer and inventor, and his team announced the development of a new process to get metallic iron. It substituted coke for the use of natural gas to perform the same chemical reaction, where the carbon in the natural gas reacts with the oxygen in the iron oxide to form carbon dioxide and carbon monoxide, thus producing metallic iron in the form of pellets, called 'sponge iron'.

Mexico's main technological contribution to the world steel industry is to lower costs and improve and increase the melting processes necessary for steel production. This process, known as direct reduction, has also evolved, and currently, there are other variants based on the same reduction to metallic iron. This change in the steel industry meant an improvement concerning steel production with coke, starting by pointing out that the use of natural gas in the furnaces reaches a metallisation of 93%, energy consumption as natural gas is 2.20–2.35 gigacalories/tonne (Gcal/tonne), and electrical energy uptake from 60 to 80 kWh/tonne, thus decreasing the investment cost of the steel complex, lowering the cost of liquid steel production to around US$8/tonne, and increasing furnace production by 2–5%, compared to coke feeding, and now it is free of sulphur and ash (Flores et al. 2001).

Since Hojalata and Lámina SA (HYLSA by its initials in Spanish; Tinplate and Sheet SA) was established in Mexico, some pioneering efforts in steel production have made contributions of great importance not only for the country but also for the steel industry in general, which have also influenced the environment. For example, no other steel company has more experience than HYLSA in direct reducing furnace (DRF) melting. Its contributions are summarised in Table 2.

In about five decades, Mexico has contributed significantly to technological progress in steel production and thus to energy savings and environmental mitigation by reducing atmospheric emissions and saving natural resources, such as using natural gas in the previously wasted process. However, neoliberal policies from 1990 to 2019 have promoted 26,559 mining concessions, causing approximately 400 social and environmental impacts, including exploiting 106 national aquifers (Monroy-Ortiz and Escobar Salazar 2019).

In terms of the steel industry, 39% (7 metric tonne of carbon dioxide equivalent [$MtCO_2e$]) of this reduction is due to energy efficiency in the industry. Another 11% (2

$MtCO_2e$) corresponds to the change of technology from AH-CBO to RDH-HAE, because it has natural gas as its primary fuel. Innovations in the steel industry to transform it into a green and environment friendly industry to reduce its impacts depend on technological advancement and patent registration (Organisation for Economic Cooperation and Development (OECD 2015).

In addition, with a significant reduction in the percentage of SO_x and NO_x emissions considering the 1957 production line at the 1979 technology change (natural gas injection), improvement in efficiency reduced these emissions by 10%, and according to Mc Kinsey in Blenkinsop (2022) estimates for CO_2, the global reduction potential for the industrial sector was approximately 19 $MtCO_2e$ by 2020 but points out that as long as coal is not replaced, the achievements would be on a low scale.

Steel plants require 28.1 m^3 of water/tonne of liquid steel that is discharged into rivers, streams, and lakes, and by vaporisation, 3.3 m^3 of additional volume is emitted to the atmosphere during steel cooling (Gómez 2002). Nowadays, as part of energy efficiency, the steam is reintroduced into the process as a motive power to drive rolling mills, improving energy use. An international comparative study on CO_2 emissions in iron and steel production in Germany, China, the United States, and Mexico showed that Mexico has the lowest CO_2 emissions, mainly due to the use of EAFs, which are less energy-intensive than blast furnaces and therefore require less energy (Hasanbeigi et al. 2016).

The steel industry has developed technology throughout the century to increase production, reduce costs, and improve the quality of steel. However, it has impacted the environment simultaneously, including generating between 220 and 421 kg of slag per tonne of iron and steel manufactured in blast furnaces (Rondón Quintana et al. 2018). It was detected from hundreds of tonne of environmental liabilities in the last century that they contained a low level of iron, calcium oxides, sodium, magnesium, and silicon through technology. Now they are used safely as they are processed to obtain sand, stone, and cement. According to Cámara Nacional de la Industria del Hierro y del Acero (CANACERO 2016), just over 820,000 tonnes of slag was used in cement production and construction in Mexico.

The effect of high carbon content in the direct-reduced iron (DRI) leads to reduced production costs as well as improved plant productivity due to the following reasons: the need to feed graphite or any other carbon source is avoided, and the decomposition of Fe_3C molecule into carbon and iron improves the thermal efficiency of the process because it is an exothermic reaction. This decreases the total electrical energy consumption and the time between casts in EAF. By increasing carbon content in DRI from 1.8 to 4.3%, the production cost of liquid steel decreases by about US$8/tonne and increases furnace productivity by 2%.

Concerning the benefits of high-temperature DRI, production costs decrease, but the primary use is associated with much higher furnace productivity. Owing to the sensitive energy input made by DRF, electricity consumption, electrode and refractory consumption, and time between casts in EAF decrease significantly.

4 Discussion and Conclusions

According to the bibliographic review, Mexico has contributed significantly to the technological progress of steel production and thus to energy savings and environmental mitigation by reducing atmospheric emissions and saving natural resources, such as using natural gas, a previously wasted resource.

As observed, the development of steel as well as human development has adverse effects, which involve tremendous environmental impacts generated by production processes to produce steel. Most of these emissions into the atmosphere and earth are the results of burning fossil fuels to power machines and manufacturing plants that process steel and manufacture various products that use it as a raw material.

All these have fostered steel production from the most incipient bases of the fourteenth century to the present period. However, they have been substantial. Efforts are required to multiply the research work aimed at optimising the process, and the analysis of multiple technical and environmental variables that interact in the production of steel and extraction of mineral resources to provide tools related to environmental protection, use of alternative energies, and climate changes for decision-makers to make the steel industry a sustainable industry that continues to operate for the future generations.

One challenge for the Mexican steel industry is to take a life-cycle view of the product obtained from this material. Another challenge is to improve efficiency in using resources and energy to produce steel and derivatives at each production stage. It is also necessary to take advantage of by-products and waste. Some of the sustainable actions that can be implemented in the steel industry is to bring all key stakeholders together to create working routes to reduce environmental impacts.

References

Arai, S. (2021). Genetic link between podiform chromitites in the mantle and stratiform chromitites in the crust: A hypothesis. *Minerals*, **11(2):** 1–10. https://doi.org/10.3390/min11020209

Ávila Juárez, J.Ó. (2011). Acero, nacionalismo y neoliberalismo en México. Universidad Autónoma de Querétaro (UAQ), Santiago de Querétaro, Qro., Mexico. Available at: https://www.scielo.org.mx/scielo.php?script=sciarttext&pid=S0188-28722013000100009

Aziz, I.H., Abdullah, M.M.A.B., Salleh, M.A.A.M., Ming, L.Y., Li, L.Y., Sandu, A.V., Vizureanu, P., Nemes, O. and Mahdi, S.N. (2022). Recent developments in steelmaking industry and potential alkali activated based steel waste: A comprehensive review. *Materials*, **15(5):** https://doi.org/10.3390/ma15051948

Black, J. and Kohser, R.A. (2008). Ferrous metals and alloys. *In:* L. Ratts (Ed.), Materials & Processes in Manufacturing, 10th edn., pp. 118–138. John Wiley & Sons. https://mmsallaboutmetallurgy.com/wp-content/uploads/2019/02/materials-processes-and-manufacturing.pdf

Bolis-Morales, G. (2017). El hierro al inicio de la Nueva España. *Boletín De Monumentos Históricos*, **35:** 70–99. https://revistas.inah.gob.mx/index.php/boletinmonumentos/article/view/10972%0A

Cámara Nacional de la Industria del Hierro y del Acero (CANACERO). (2016). Residuos y subproductos. Available at: https://www.canacero.org.mx/aceroenmexico/descargas/residuos_2018.pdf

Cortez-Lugo, M., Riojas-Rodríguez, H., Moreno-Macías, H., Montes, S., Rodríguez-Agudelo, Y., Hernández-Bonilla, D., Catalán-Vázquez, M., Díaz-Godoy, R. and Rodríguez-Dozal, S. (2018). Evaluation of the effect of an environmental management program on exposure to manganese in a mining zone in Mexico. *NeuroToxicology*, **64:** 142–151. https://doi.org/10.1016/j.neuro.2017.08.014

Equihua, F., Salinas, A. and Nava, E. (2010). Effect of aluminum content on austenite-ferrite transformation temperature in low carbon (Si-Al) hot rolled GNO electrical steels. Revista de Metalurgia (Madrid), **46(1):** 78–85. https://doi.org/10.3989/revmetalm.0938

Flores, S., Rocio, M. del. and Parellada, F.S. (2001). Cambio tecnológico en la industria siderúrgica Mexicana. *El Cotidiano*, **21(109):** 97–106. https://www.redalyc.org/articulo.oa?id=32510911

Giacalone, R. (2010). Análisis comparado de las trayectorias estratégicas de empresas siderúrgicas latinoamericanas. Desarrollo interno y contexto histórico. *Investigaciones de Historia Económica*, **6(16):** 43–72. https://doi.org/10.1016/s1698-6989(10)70034-5

Gómez, L.M. (2002). Acero. La Ciencia/80 para Todos (3rd ed.). Fondo de Cultura Economica (Fondo), Mexico.

Han, C., Xiao, W., Zhao, G., Sun, M., Qu, W. and Du, A. (2014). Re-Os geochronology on molybdenites from the Donggebi Mo deposit in the Eastern Tianshan of the Central Asia orogenic belt and its geological significance. *Resource Geology*, **64(2)**: 136–148. https://doi.org/10.1111/rge.12033

Hasanbeigi, A., Arens, M., Cardenas, J.C.R., Price, L. and Triolo, R. (2016). Comparison of carbon dioxide emissions intensity of steel production in China, Germany, Mexico, and the United States. Resources, Conservation and Recycling, **113:** 127–139. https://doi.org/10.1016/j.resconrec.2016.06.008

Hobsbawm, E.J. (2003). En Torno a los Orígenes de la Revolución Industrial (Sexta). Siglo XXI de España Editores, Madrid, Spain (SA).

Kalpakjian, S. and Schmid, S.R. (2010). Ferrous metals and alloys: Production, general properties, and applications. *In:* Manufacturing Engineering and Technology (xxx), 6, pp. 129–150. Pearson, London.

Moe, K.S., Yang, J.S., Johnson, P., Xu, X. and Wang, W. (2018). Spectroscopic analysis of microdiamonds in ophiolitic chromitite and peridotite. *Lithosphere*, **10(1):** 133–141. https://doi.org/10.1130/L603.1

Moffit, A. (1998). Industrias basadas en recursos naturales. Hierro y acero. *In:* Enciclopedia de Salud y Seguridad en el Trabajo (xxxx), pp. 73–93. https://www.insst.es/documents/94886/161958/Sumario+del+Volumen+I.pdf/18ea3013-6f64-4997-88a1-0aadd719faac?t=1526457520818

Monroy-Ortiz, R. and Escobar Salazar, J.A. (2019). Mecanismos de política pública para el emplazamiento de la minería en México. Procesos e impactos. *Revista de Direito Da Cidade*, **11(1):** 139–160. https://doi.org/10.12957/rdc.2019.33764

Organisation for Economic Cooperation and Development (OECD). (2015). Greening steel: Innovation for climate change mitigation in the steel sector. Available at: https://www.oecd.org/sti/ind/Environmental-patents-steel.pdf.

Pandiand, M.S. and Varma, O.P. (2001). Geology and geochemistry of topaz granite and associated wolframite deposit at Degana, Rajasthan. *Journal Geological Society of India*, **57:** 297–307.

Pérez Sánchez, B., Mayo Castro, A. and Torrez Vázquez, J. (2015). Evolución histórica de alfa: Un grupo económico de capital nacional. *Hitos de Ciencias Económico Administrativas*, **21(59):** 19–32. https://doi.org/10.19136/hitos.a0n59.1014

Robledo Monsivais, R. (1997). Analisis metalográfico y termoquímico de aceros y fundiciones. Universidad Autonóma de Nuevo León, San Nicolás de los Garza, Mexico. Available at: Chrome-extension://efaidnbmnnnibpcajpcglclefindmkaj/http://cdigital.dgb.uanl.mx/te/1020119025/1020119025_02.pdf.

Rondón Quintana, H.A., Muniz de Farias, M. and Reyes Lizcano, F.A. (2018). Uso de escorias de alto horno y acero en mezclas asfálticas: revisión. *Revista Ingenierías Universidad de Medellín*, **17(33):** 71–97. https://doi.org/10.22395/rium.v17n33a4

Salcedo Sánchez, E.R., Esquivel Martínez, J.M., Morales, M.M., Talavera Mendoza, O. and EstellerAlberich, M.V. (2022). Ecological and health risk assessment of potential toxic elements from a mining area (water and sediments): The San Juan-Taxco River system, Guerrero, Mexico. Water (Switzerland), **14(4):** xxx. https://doi.org/10.3390/w14040518

Sánchez Díaz, G. (2009). Los orígenes de la industria siderúrgica mexicana. Continuidades y cambios tecnológicos en el siglo XIX. *Revista de Estudios Históricos*, **50:** 11–60. http://www.redalyc.org/articulo.oa?id=89814806001

Schiavo, B., Morton-Bermea, O., Salgado-Martínez, E., García-Martínez, R. and Hernández-Álvarez, E. (2022). Health risk assessment of gaseous elemental mercury (GEM) in Mexico City. *Environmental Monitoring and Assessment*, **194(7):** 1-19. https://doi.org/10.1007/s10661-022-10107-7

Soriano V.M.C. (2010). La huerta del Colegio de San Gregorio, asiento del taller de Manuel Tolsá y su transformación en fundición de cañones, 1796–1815. *Historia Mexicana*, **59(4):** 1401–1418. https://historiamexicana.colmex.mx/index.php/RHM/article/view/1838

Toledo Beltrán, D. and Zapata, F. (1999). Acero y Estado: Una historia de la industria siderúrgica integrada de México (D. de C. S. y Humanidades, ed.). Universidad Autónoma Metropolitana, Unidad Iztapalapa, Ciudad de México, Mexico.

Wei, C., Huang, Z., Yan, Z., Hu, Y. and Ye, L. (2018). Trace element contents in sphalerite from the nayongzhi Zn-Pb deposit, northwestern Guizhou, China: Insights into incorporation mechanisms, metallogenic temperature and ore genesis. *Minerals*, **8(11):** 1-23. https://doi.org/10.3390/min8110490

Wurl, J., Lamadrid, M.I., Mendez-Rodriguez, L. and Vargas, B.A. (2018). Arsenic concentration in the surface water of a former mining area: The la junta creek, bajacalifornia sur, Mexico. *International Journal of Environmental Research and Public Health*, **15(3):** 1-22. https://doi.org/10.3390/ijerph15030437

Zhang, Y., Sun, S., Xu, C., Wang, H., Zhao, C. and Zhu, Z. (2003). Carbonatitic volcanic genesis of Hetaoqing Fe-Cu deposit in central Yunnan, China. Resource Geology, **53(4):** 261–272. https://doi.org/10.1111/j.1751-3928.2003.tb00175.x

Zhang, J., Huang, W., Wu, J., Liang, H. and Lin, S. (2021). Geological implications of apatite within the granite-related Jiepai W–(Cu) deposit, Guangxi, South China. *Ore Geology Reviews*, **139(PB):** 104548. https://doi.org/10.1016/j.oregeorev.2021.104548

Geological Effects on Quarry Blasting Fragmentation—Study on Limestone Mining of Southern India

Shanmugasundaram Arumugam, Sivakumar Karthikeyan, Saubhagya Ranjan Mahapatra, Venkatramanan Senapathi, Jenefer Sebasthiyan, Ramachandran Annadurai, and Mohammad Gholizadeh

1 Introduction

Mining is an important part of humanity's existence, as it gives rise to industrial growth, ancillary industries, job creation, and wealth (Keyser et al. 2020). India produces 95 minerals, including four fuel-related minerals, ten metallic minerals, 23 non-metallic minerals, three atomic minerals, and 55 minor minerals (Krishna et al. 2020). Growth in the industry is being driven by increased infrastructural development, car manufacturing, the power and cement sectors, and the residential and commercial building industries. The Indian mining sector, which uses non-renewable resources to meet societal demands, contributes significantly to the society and the nation's growth. Every 1% rise in the growth rate of the mining sector leads to a 1.2–1.4% increase in industrial output and a 0.3% increase in India's gross domestic product (GDP) (Anon 2018). In addition, it directly employs over 2.5 million people. India is one of the top ten producers of major metallic and non-metallic materials, such as iron ore, bauxite, coal, limestone, and manganese. The mining sector has made important contributions to the nation's infrastructural development and catalysed vast economic growth (Mishra and Swain 2020). However, the reliability of procedures employed to bore holes is poor. To improve breakage, drillers should be trained to keep records indicating the depth at which the drill bit enters and exits each rock encountered. These techniques will give a chance to enhance charge placement, decking placement, and stemming materials.

The development and expansion of open-pit mines with greater output ambitions have been achieved by the use of large-capacity earth moving equipment, continuous mining equipment, enhanced rock-blasting technology, process innovation, and increasing use of information and computational technologies (Osanloo and Paricheh 2020). However, the performance of these types of equipment is heavily impacted by blast performance, specifically fragment size, distribution, and muck profile. Embedded boulders in mining can cause equipment damage, safety hazards, production delays, and cost overruns. The presence of embedded boulders on the surface benches of worn limestone quarries is widespread in nature. Spheroidal weathering on fractured rock creates embedded boulders, which are thrown out in the muck pile without breaking during the blasting operation (Lak et al. 2019). However, identifying such in-situ boulders in a geological setting is a difficult task and the reliability of such identification systems is poor. As a result, huge blocks requiring secondary breaking or an excess of particles might occur because of poorly constructed blasts or unfavourable geological circumstances (Bui et al. 2020). The authors concluded that a case study of a limestone mine with unique geological features, such as embedded boulders covered in soft soil, necessitates changes in existing blast design parameters and the implementation of correct blasting practices.

S. Arumugam
Department of Geology, Malankara Catholic College, Manonmanium Sundaranar University, Kanyakumari, India

S. Karthikeyan (✉)
Department of Geology, Faculty of Science, Alagappa University Karaikudi, Karaikudi, India
e-mail: siva.karthi90@yahoo.com

S. R. Mahapatra
Department of Petroleum Engineering, Chandigarh University Mohali, Mohali, India

V. Senapathi · J. Sebasthiyan
Department of Geology, National College, Tiruchirappalli, India

R. Annadurai
Department of Applied Geology, University of Madras, Chennai, India

M. Gholizadeh
Department of Fisheries, Gonbad Kavous University Gonbad Kavous, Gonbad-e Kavus, Iran

J. Marrugo-Negrete et al. (eds.), *Remediation and Carbon Capture in Mining Sector*, https://doi.org/10.1007/978-3-031-87883-1_2

1.1 Occurrence

Boulder is a corestone created by spheroidal weathering on fractured bedrock surrounded by concentric ringlet layers and saprolite (Silveira et al. 2022). It is also known as a spherical or cylindrical form, has varying diameters, and may be found wherever in the geologic material that it encounters (Yang et al. 2021). A boulder is also described as a huge rounded lump of rock with a diameter larger than 0.3 m that rests on the ground's surface or embedded in the silt and soil (Gangrade et al. 2022). The form of the boulder ranges from spherical to ellipsoidal, with some being almost perfect spheres and others being nearly cubic with rounded edges (Watkins et al. 2019). The varied boulder shapes are caused by the spheroidal weathering process of fractured rock, which transforms the polyhedron shape of the rock into a spherical shape. The spheroidal process gradually reduces the volume of the boulder, and as it becomes rounded, the rate of weathering is slower than in other shapes with many edges and corners (Jamtveit and Hammer 2011). These rocks usually consist of three to six concentric sheets or layers around them. The concentric layers of the boulder have been given numerous names by experts, including layers or onion-skin layers, shells or spherical shells, and rindlets. Some geological and petrographic criteria, such as the degree of interlocking in the rock micro fabric or the quality of binder minerals (e.g. in sandstone), may be examined qualitatively, but only very rarely in categories such as weathering or alteration phases in a rock mass (IAEG 1981; ISRM 1978). Most rock formations in the studied region undergo weathering, which is a natural process that causes a new rock to become entirely weathered and residual soil. Weathering in metamorphic rocks can occur as deep as 50 m in thickness; however, it is most often weathered up to 10–20 m in most situations. Furthermore, the high intensity of rainfall in the tropical area would hasten the response of weathering on a specific rock mass, whether on the ground surface or embedded in the soil stratum. The response of spherical weathering on the broken rock block progressively turns intact bedrock into a chemically weathered rock by penetration of rainwater deep into bedrock under the soil layer. Deep weathering of a broken rock block causes several concentric cracks on the rock, gradually transforming the rock into an oblong or rounded shape, known as a boulder.

1.2 Identification

The bulk of spots on the west, east, and northern sides of the band have exploratory open pits, and their continuity can be seen all the way to the base of the hill. These steps are typically used in quarries or mines to find the location and characteristics of such embedded boulders. Each strategy, though, has benefits and drawbacks. Based on the features of the site and the requirements of the research, a workable method for pinpointing the location of such boulders can be used. The creation of structural mapping (Fig. 1) is the traditional

Fig. 1 Geological mapping for embedded boulders in limestone benches

method for locating embedded boulders, but it may not be a practical way to locate such a formation, as it heavily depends on the geologists' knowledge of the area. The ground-penetrating radar is a useful tool that can accurately locate underground storage tanks, utilities, buried drums, hazardous materials, and even items beneath reinforced concrete floors and slabs, both metallic and non-metallic (Quinta-Ferreira 2019). This radar can also be used to map trenches and excavations and to detect pollutant plumes in some instances. Additionally, it has been applied in archaeology to discover buried artifacts. However, *ground-penetrating* radar (GPR) data can be challenging to interpret due to reflections caused by boulders and debris, which can resemble pipelines and tanks, and subtle variations in electrical properties generated by differences in soil type, mineralogy, grain size and moisture content, producing 'noise' that is difficult to decipher (Dai et al. 2022). Thus, analysing radargrams is both scientific and artistic endeavour. Surface geophysical technologies, such as direct current (DC) resistivity and electrical resistivity tomography (ERT), involve passing an electric current into the ground through two electrodes and measuring surface voltages to determine the direction and magnitude of current flow in the subsurface (Acosta et al. 2022). By comparing the resistivity of materials in the subsurface with a standard chart provided by geologists, we can create a subsurface map and identify any deposits (Karthikeyan et al. 2014). In addition, exploratory drilling operations can determine the location of embedded boulders through drill hole logging. This method involves examining the variety of minerals present at different depths to gain insight into the subsurface. Drill chips generated during drilling can also help analyse the position of boulders. However, the skill and competence of drilling machine operator and sample crew are critical factors in this process.

1.3 Influences in Blasting Operations

The creation of embedded boulders from blasting always makes it difficult to operate the loading equipment properly and causes crusher obstruction if the mineral is valued. Boulders are created at a typical explosion site due to the presence of hard and large rock in the stemming zone (uncharged area), bedding plane slabs from inside the blast, hard rock at or near intrusive dykes, and rock isolated from explosive energy due to severe back break (Jimeno et al. 1995). If the embedded un-fissured boulders or floaters are encountered, the strain waves generated by the explosion travel with minimal attenuation in the floaters, but their energy is swiftly dissipated in the matrix. Because imbedded boulders that do not carry any explosive charge receive relatively little strain wave energy, they are often hurled out 'intact' into the blasted muck pile (Fig. 2). The phrase 'embedded boulders' is also used to describe 'floaters', which are un-fissured boulders of hard rock embedded inside a considerably softer/weaker matrix (Hagan and Reid 1983). The features of embedded boulders differ from those of conventional boulders as well as those of the surrounding rock mass. Generally, these boulders are not fractured during blasting and instead erupt out of the soil or any other matrix that causes the boulders to emerge. Normal boulders, on the other hand, are produced by blasting operations due to the existence of joints or discontinuities, faulty blast design, inadequate explosive energy, and bad blasting methods. Hence, investigating the nature and occurrence of embedded boulders is necessary to poise the excavation process in a cost-effective manner. The embedded boulders are situated just above the blasting material and which is affected by various geological agents. The objective of the present study is to understand the geological effects

Fig. 2 Study area map

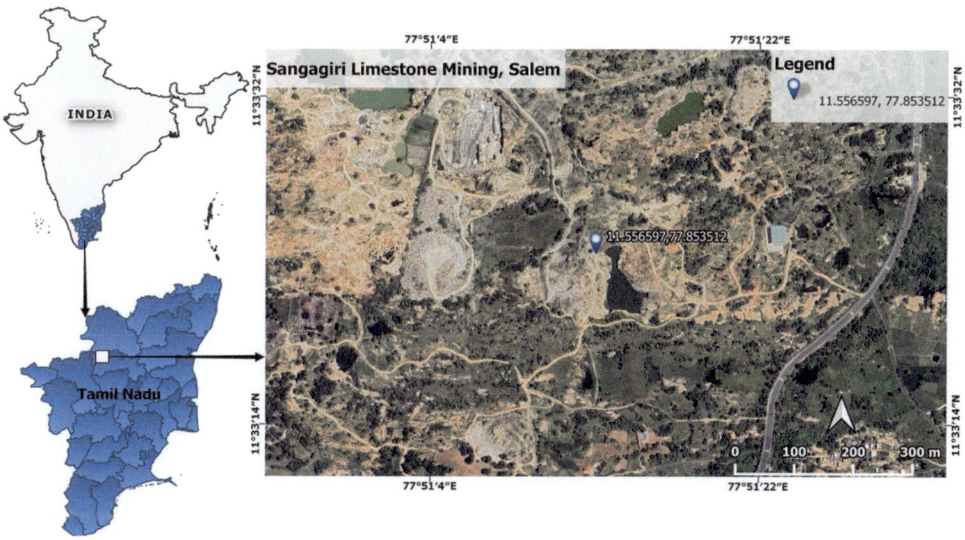

of blasting operations and fragmentations where embedded boulders occur.

2 Site Description for Case Study

The research area includes a limestone quarry with a mining lease area of 21.65 ha, which is located in the northern region of the Indian state of Tamil Nadu (Fig. 2). The overall strike length of 1.5 km is split into a cluster of tiny pits for operating flexibility. The depth of these trenches ranges between 20 and 37 m below the ground level. The metamorphic rock contains limestone with a density of 2.6 g/cc. The mine is worked in benches, with the top soil cover of black cotton soil dug (without blasting) using diesel-powered hydraulic excavators of 3.6 cum and haul trucks of 38-ton capacity. However, the successive benches of limestone strata were dug by the traditional way of breaking, which is drilling and blasting operations.

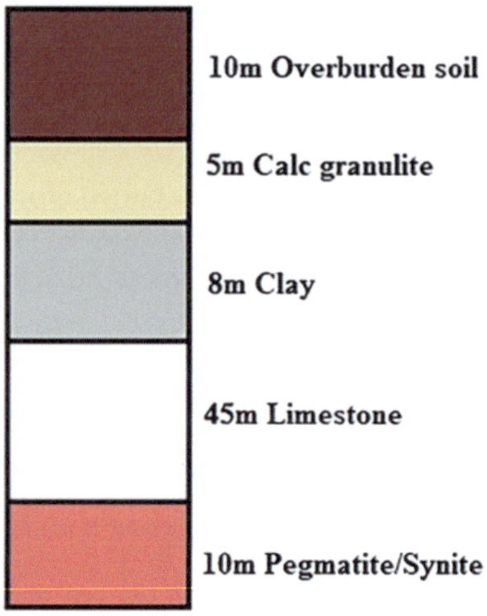

Fig. 4 Drill hole logging

2.1 Geology

Peninsular gneiss is the oldest rock formation, with enormous granite gneiss overlaid by a rich accumulation of recent quaternary formation (GSI 1995, 1998). On a regional scale, the granite gneiss is N40°E–S40°W with a SE70° dip. Massive calc continuous band granulites form individual, hundreds of meters thick layers that are laterally inter-banded with crystalline limestones and garnet-biotite-sillimanite graphite gneisses (Ganguli et al. 2021). Its notable characteristics are extreme fracturing (Fig. 3) and fine-grained texture. The region receives around 998 mm of rainfall every year, with the wet season lasting mostly from October to January during the northeast monsoon. Summers are hot, with maximum temperatures reaching 42 °C, while winters have low temperature of 21 °C. Earlier investigations by other authors (Nageswara Rao and Srinivasan 1980) also suggest that the geology of the Palaghat region is dominated by Precambrian metamorphic complex rocks, with sediments forming a thin strip along the coastal plains. The extant gap in the formation is flanked to the north by khondalite, related calc granulites, and crystalline limestones, and is formed primarily of migmatitic gneisses and associated granites. The diversity of lithology is identified and the corresponding bench face is illustrated in Fig. 4, which is based on borehole logging performed following exploratory drilling. Each rock necessitates a specific drill and blast design depending on its features

Fig. 3 Geology map

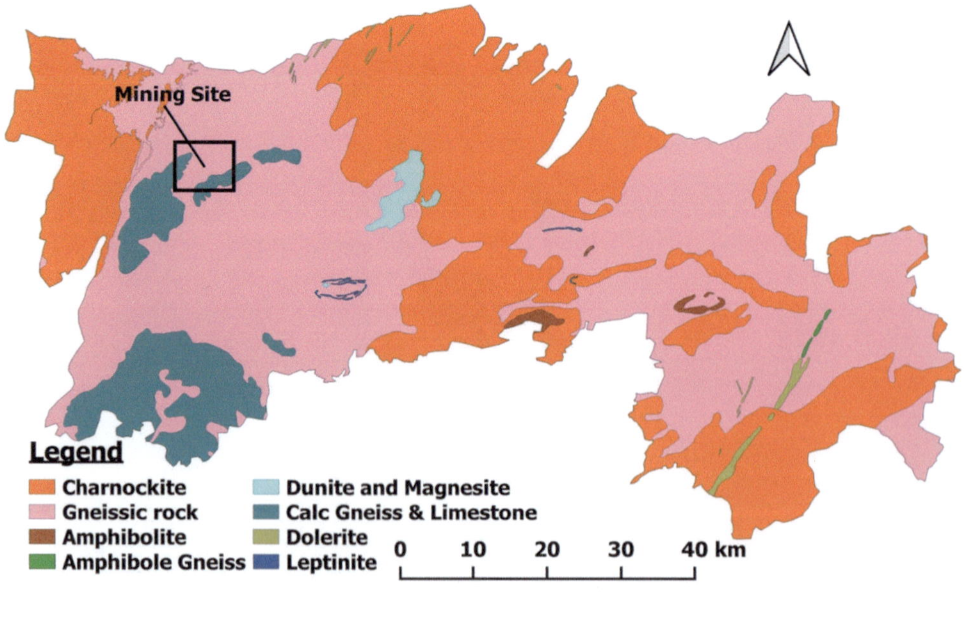

in order to maximise fragmentation while minimising environmental hazards, such as ground vibration, fly rock and so on (Sasaoka et al. 2011).

3 Methodology

3.1 Quarrying Activities

The quarrying of limestone in the examined mines is done using traditional open-pit mining methods, which include blasting activities. The bench blasting is carried out using hydraulic down-the-hole (DTH) rotary-percussive apparatus with a drill hole of 115 mm diameter. During the blasting process, the detonating cord and NONEL system are employed for commencement (Rasagna et al. 2021; Bakhtavar et al. 2022). The blasted muck is dug by a 3.6-cum diesel-driven hydraulic excavator and loaded into dump trucks of 38-ton capacity prior to being hauled to the crusher or waste dump on the quarry leasehold area's surface (Fig. 5). The rock breaker is used for the subsequent shattering of coarser material generated by blasting.

3.2 Fragmentation Evaluation

To investigate the effect of fragmentation of embedded boulders, three trial explosions were conducted in the identical formation using various blast design parameters and similar initiation patterns, namely, echelon pattern and open chevron. The NONEL shock tube starting system was employed in all blasts to ensure successful initiation and improved breakage. The Digital Image Analysis Technique (DIAT) program 'Fragalyst 4.0' was used to estimate the fragmentation quality of the blasted muck pile by determining fragmentation distributions resulting from blasting of different rock types, describing the rocks, and optimising blasting (Balamadeswaran et al. 2017a). To avoid or limit the impact of difficulties during recording, several fundamental measures must be performed in order to obtain trustworthy findings in image analysis (Raina et al. 2012).

3.3 Blast Experiments

To determine the effects of geological conditions and present blasting procedures, a series of three experimental blasts were conducted at the aforementioned quarry to analyse fragmentation size distribution in the upper benches, where embedded boulder development is typical. Decking procedures do not incorporate blast design. About 40–50 photographs of shattered muck piles per blast were captured in adequate lighting and after 30–45 min to account for swelling. The analysis in the Fragalyst software is then used to obtain the fragment size distribution. Table 1 shows the blast design parameters for the experimental explosion. The system displays graphical findings in the form of various distributions, such as the Rosin–Rammler curve and normal distributions (Fig. 6). Photographs of the blasted muck pile were taken with a reference scale item, such as a ball of known diameter, and were compared with the standard photograph to determine blast fragmentation (Akinbinu et al. 2022).

Fig. 5 Quarry activities with fragmentation of boulders on the surface

Table 1 Blast design and trial blast performance parameters

Parameters	Blast no. 1	Blast no. 2	Blast no. 3
No. of holes	6	7	6
Depth of the hole (m)	4	5	7
Spacing and burden (m)	3 and 2.5	3 and 2.5	3 and 2.5
Drilling pattern	Staggered	Staggered	Square
Stemming length (m)	0.8	1.0	2.0
Column charge	ANFO	Slurry	ANFO
Primer charge	Slurry	Slurry	Slurry
% of Primer	20.40	18.30	42.60
Max. charge/delay (kg)	26.61	17.06	70
Initiation system	NONEL	NONEL	NONEL
Volume of material (m3)	331.15	313.38	1192.80
Charge factor (kg/m3)	0.98	0.50	0.89
Spacing/burden ratio	1.2	1.2	1.2
Blast hole dia/BH ratio	32.55	26.85	12.62
Stiffness ratio	1.6	1.8	3.4
Initiation pattern	V-pattern	Diagonal	Row-by-row
Mean fragment size (m)	0.26	0.68	0.586

4 Outcomes and Discussions

Locating embedded boulders and drilling in such areas is challenging, and as a result, it has been discovered that using the 'V'-initiation pattern of firing explosives leads to better fragmentation and larger fragment sizes, compared with the diagonal line of fire or row-by-row line of firing in the same rock environment. This is likely due to the presence of embedded boulders in the soft rock matrix, which causes the explosive energy to escape from the blast hole via the soil, resulting in suboptimal fragmentation. The fragment size was observed to increase linearly with the stiffness ratio, and a stiffness ratio of 3:4 is considered optimal for fragmentation. However, the explosions studied did not achieve optimal fragmentation, instead, the particle size increased dramatically, leading to the formation of boulders (Fig. 7). This might be due to the explosive charge column not making contact with the embedded pebbles, and a shorter stemming length, which was not maintained in accordance with the design, thus increasing the likelihood of explosive energy being released without being used. The type of material used for stemming also plays a crucial role in determining fragmentation level, and the use of fine drill cuttings has the potential to cause poor fragmentation and flyrock (Balamadeswaran et al. 2017b; Rasagna et al. 2021). In the presence of a clay cover over embedded boulders, the row-by-row or diagonal line of fire generated fly rock to a distance of over 200 m during blasting whereas the 'V' line of firing resulted in a lower fly rock distance of less than 200 m.

5 Conclusions

The blast-monitoring procedure is now underway, with site supervision provided by the mine geology team and assistance from the drill and blast crew. The management operating system (MOS) is a control, communication, and activity system that has been created to meet corporate goals and objectives in a timely and effective manner. Geological investigations must be carried out to determine the existence of weakness planes, discontinuities, voids, and fractures in order to achieve optimal rock fragmentation. Studies have already been done on the many facets of fragmentation, but blasting in benches made up of embedded boulders buried in soft soil is a different kettle of fish. To improve fragmentation when blasting in such formations, the bench face must be geologically evaluated using appropriate techniques based on site-specific conditions. Effective communication with the mining geologist is key to improving blasting operation quality. To identify embedded boulders in soft rock matrix, holes should be bored in a staggered pattern, charged with explosives with

Fig. 6 Rosin–Rammler curve

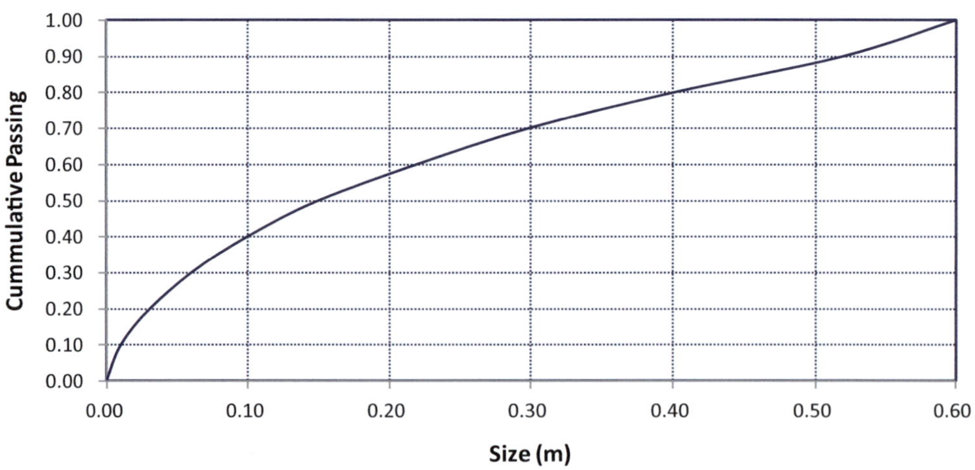

Fig. 7 Ineffective fragmentation

high shock energy, and triggered using 'V' patterns with a suitable delay interval. In the absence of any equipment for identifying the buried boulders, stab holes of shorter length and smaller diameter can be bored in-between the standard planned holes. To improve breakage, drillers should be trained to keep records of the depth at which the drill bit enters and exits each rock encountered.

References

Acosta, J.A., Gabarrón, M., Martínez-Segura, M., Martínez-Martínez, S., Faz, Á., Pérez-Pastor, A., Gómez-López, M. D., et al. (2022). Soil water content prediction using electrical resistivity tomography (ERT) in Mediterranean tree orchard soils. *Sensors*, **22(4):** 1365.

Akinbinu, V.A., Oniyide, G.O. and Idris, M.A. (2022). Assessment of rock fragmentation and strength properties using the Rosin-Rammler and Extended Swebrec Distribution functions parameters. *International Journal of Mining and Geo-Engineering*, **56(1):** 53–60.

Anon. (2018). Metals & Mining Industry in India. India Brand Equity Foundation, New Delhi, India.

Bakhtavar, E., Abdollahisharif, J. and Mohammadi, D. (2022). Analysis and improvement of blasting operation in porphyry, diorite dyke, and trachyte Sungun zones: In-situ investigations. *International Journal of Mining and Geo-Engineering*, **56(1):** 19–24.

Balamadeswaran, P., Mishra, A.K., Sen, P. and Ifthikhar Ahmed, M. (2017a). Blast performance analysis using digital image processing technique—Key to unlock productivity in quarries. *Journal of Mines, Metals and Fuels*, **65(5):** 245–251.

Balamadeswaran, P., Mishra, A.K., Singh, D.R. and Sen, P. (2017b). Effect of blast induced fragmentation on performance of hydraulic excavator: A case study. *NexGen Technologies for Mining and Fuel Industries*, **1 & 2**: 333.

Bui, X.N., Nguyen, H., Le, H.A., Bui, H.B. and Do, N.H. (2020). Prediction of blast-induced air over-pressure in open-pit mine: Assessment of different artificial intelligence techniques. *Natural Resources Research*, **29(2):** 571–591.

Dai, Z., Zhang, M. and Gu, J. (2022). GPR forward modelling and imaging analysis for boulder detection. *Iranian Journal of Science and Technology, Transactions of Civil Engineering*, **46(4):** 3193–3202.

Gangrade, R.M., Grasmick, J.G. and Mooney, M.A. (2022). Probabilistic assessment of void risk and grouting volume for tunneling applications. *Rock Mechanics and Rock Engineering*, **55**(5): 2771–2786.

Ganguli, S.S., Pal, S.K., Sundaralingam, K. and Kumar, P. (2021). Insights into the crustal architecture from the analysis of gravity and magnetic data across Salem-Attur Shear Zone (SASZ), Southern Granulite Terrane (SGT), India: An evidence of accretional tectonics. *Episodes Journal of International Geoscience*, **44(4):** 419–442.

GSI. (1995). Geological Quadrangle Map of 58/I. Geological Survey of India, Kolkata, India.

GSI. (1998), Geological Map of India on 1:2,000,000 Scale. Geological Survey of India, Kolkata, India.

Hagan, T.N and Reid, I.W. (1983). Performance monitoring of production blast hole drilling—A mean of increasing blasting efficiency. *In:* Proceedings of 2nd Surface Mining and Quarrying Symposium, Bristol, UK, Publisher: Institution of Mining Metallurgy, London, pp. 20–30.

IAEG—International Association of Engineering Geology (Ed.). (1981). Rock and soil description and classification for engineering geological mapping. Report by the IAEG Commission on Engineering Geological Mapping. *Bulletin of Engineering Geology and the Environment*, **24**: 235–274.

Ifthikar Ahmed, M., Balamadeswaran, P., Ajay Kumar, L. and Palanikumaresan, M. (2014), Role of structural geology in surface mine blasting, Proceedings of the 6th National Seminar on "Surface Mining" (NSSM), January 10~11, Indian School of Mines Dhanbad, pp. 111–118.

ISRM—International Society for Rock Mechanics (Ed.). (1978). Suggested methods for the quantitative description of discontinuities in rock masses. Commission on Standardization of Laboratory and Field Tests, Document No. 4. Abstracts. *International Journal of Rock Mechanics and Mining Sciences & Geomechanics*, **15**: 319–368.

Jamtveit, B. and Hammer, Ø. (2011). Chapter 7: Hierarchical fracturing during weathering and serpentinisation. *Geochemical Perspectives, European Association of Geochemistry, Houten, Netherland*, **1(3):** 418–432.

Jimeno, C.L., Jimeno, E.L. and Carcedo, F.J.A. (1995). Drilling and Blasting of Rocks. A.A. Balkema, Rotterdam, the Netherlands, 391 p.

Karthikeyan, E., Sakthivel, P., Sarath Prasana, K.B., Balamadeswaran, P. and Magesh, G.R. (2014). Detection and mitigation of boulder formation in surface mine blasting. *In:* Proceedings of Expo and Symposium on Mining (MineFest India, 2014), Neyveli Mining Engineers Association of India (Tamil Nadu Chapter), pp.185–195.

Keyser, E., Adeoluwa, A.S. and Fourie, R. (2020). Time pressure, life satisfaction and absenteeism of employees of shift work within the mining industry in South Africa. *International Journal of Social Sciences and Humanity Studies*, **12(2):** 255–272.

Krishna, R.S., Mishra, J., Meher, S., Das, S.K., Mustakim, S.M. and Singh, S.K. (2020). Industrial solid waste management through sustainable green technology: Case study insights from steel and mining industry in Keonjhar, India. *Materials Today: Proceedings*, **33:** 5243–5249.

Lak, M., Marji, M.F., Bafghi, A.Y. and Abdollahipour, A. (2019). Analytical and numerical modeling of rock blasting operations using a two-dimensional elasto-dynamic Green's function. *International Journal of Rock Mechanics and Mining Sciences*, **114:** 208–217.

Mishra, D.P. and Swain, S.K. (2020). Global trends in reserves, production and utilization of iron ore and its sustainability with special emphasis to India. *Journal of Mines, Metals & Fuels,* **68:** 11–18.

Mishra, A.K., Balamadeswaran, P. and Sen, P. (2003). An approach to eco-blasting for environmentally sensitive areas—a review. *Mining Engineer's Journal*, **5(4):** 17–23.

Monjezi, M., Mohamadi, H.A., Barati, B. and Khandelwal, M. (2014). Application of soft computing in predicting rock fragmentation to reduce environmental blasting side effects. *Arabian Journal of Geosciences,* **7(2):** 505–511.

Nageswara Rao, J. and Srinivasan, R. (1980). Some Aspects of Geomorphology, Structure and Sedimentation in Palaghat Gap Area. Special Publication No. 5, Geological Survey of India, pp. 39–43.

Osanloo, M. and Paricheh, M. (2020). In-pit crushing and conveying technology in open-pit mining operations: A literature review and research agenda. *International Journal of Mining, Reclamation and Environment*, **34(6):** 430–457.

Peng, J., Zhang, F., Yan, G., Qiu, Z. and Dai, X. (2019). Experimental study on rock-like materials fragmentation by electric explosion method under high stress condition. *Powder Technology*, **356:** 750–758.

Quinta-Ferreira, M. (2019). Ground penetration radar in geotechnics. advantages and limitations. *IOP Conference Series: Earth and Environmental Science*, **221(1):** 012019. IOP.

Raina, A.K. (2012, November). A history of digital image analysis technique for blast fragmentation assessment and some Indian contributions. *Electrical Measuring Instruments and Measurements*, **5(3):** 3–11.

Rasagna, D.L.S., Reddy, K.S.N., Singh, S. and Das, S. (2021). Modified presplit blast design and its implementation to control near field blast vibrations. *Global Journal of Earth Science and Engineering*, **8:** 62–71.

Sasaoka, T., Shimada, H., Hamanaka, A. and Matsui, K. (2011). Study on blast vibration and size of fragmentation at limestone quarry. *In:* Proceedings of 20th International Symposium on Mine Planning and Equipment Selection, Almaty, Kazakhstan, Publisher; International Mining Team Publishing Ltd, England, 12–14 October 2011, pp. 714–730.

Silveira, C.S., Rodrigues, J.G., Araripe, D.R., Bernedo, A.V. and Mantovano, J.L. (2022). Geochemistry of a spheroidal weathering profile in a tropical mountainous landscape, Rio de Janeiro, Brazil. *Anais da Academia Brasileira de Ciências*, **94(Supp. 3):** e20210544.

Watkins, R.N., Jolliff, B.L., Mistick, K., Fogerty, C., Lawrence, S.J., Singer, K.N. and Ghent, R.R. (2019). Boulder distributions around young, small lunar impact craters and implications for regolith production rates and landing site safety. *Journal of Geophysical Research: Planets*, **124(11):** 2754–2771.

Yang, R., Zheng, C., Yang, L., Zuo, J., Cheng, T., Ding, C. and Li, Q. (2021). Study of two-step parallel cutting technology for deep-hole blasting in shaft excavation. *Shock and Vibration*, **2021:** 1–12.

Nickel Enrichments in Freshwater Fishes from Mining Region of Eastern Colombia: Assessing Human Health Risk

José Marrugo-Negrete, Yenis Simanca-Lozano, Siday Marrugo-Madrid, S. B. Sujitha, and M. P. Jonathan

1 Introduction

Nickel (Ni) is a pervasive naturally occurring transition element in the earth's core (6%) and crust (0.0086%) (ATSDR 2005; Topal et al. 2017) and is released into the environment through multiple sources of natural and/or anthropogenic origin (Olivares-Arias et al. 2015; Pérez-Rodríguez et al. 2014). Various anthropogenic sources namely mining, refining, smelting, electroplating, waste incineration, and production of stainless steel and Ni–Cd batteries have eventually resulted in Ni contamination worldwide (Coman et al. 2013). In recent years, rising global demand and production of Ni has resulted in adverse effects on the livelihood and environment at both local and global scales, even while improving a country's economy (Nakajima et al. 2017).

Ni is considered to be an essential trace metal in living organisms (Sun et al. 2020), where it is a major component of several enzymes that participate in the cycling of oxygen, carbon, and nitrogen ions (Ragsdale 1988; Blewett and Leonard 2017). However, at higher concentrations it can be toxic (Palermo et al. 2015), inducing lipid damage, oxidative stress, impairment of gas exchange, and inhibition of ion regulation (Chen et al. 2003; Pane et al. 2004; Leonard and Wood 2013; Kubrak et al. 2014). Ni bioaccumulation and toxicity in freshwater fishes are dependent on their bioavailability in the environment, water chemistry, and species traits (Lebrun et al. 2011). Additionally, Ni speciation plays a critical role in toxicity, as its free ion is the most bioavailable form that enters the organism through the epithelial tissues (e.g. gills) (Wood 2012). Contaminated freshwater systems can contain Ni levels of up to 100 μg/L proving to be a major pathway of Ni accumulation in freshwater fishes (EC 2008).

Colombian Scenario

Colombia with its tropical context, grassed earth, and rich biodiversity, also embraces the world's largest ferronickel mines and processing facilities on the banks of the San Jorge River in the department of Córdoba, North Colombia. Nearly 40,606 tonnes of Ni were produced and exported to China (24%), Japan (16%), Korea (13%), and India (10%) during the year 2017 (UPME 2018). Even though, Ni production shows significant long-term growth and beneficial sources of wealth, their adverse impacts on local communities are obvious through prolonged illness, miscarriages, skin abrasions, and genetic deformations.

Several studies (Marrugo-Negrete et al. 2006, 2007, 2010; Marrugo-Negrete et al. 2020a, b) have reported the alarming concentrations of heavy metals in the sediments, aquatic plants and fishes of the San Jose River Basin and the adjoining marsh complexes. Additionally, Gracia et al. (2010) determined the concentrations of metals in humans due to fish consumption from the Ayapel marsh, Cordoba, Colombia. Likewise, Cruz-Esquivel et al. (2019) determined the genotoxic damage associated with the concentrations of heavy metals in the human population that live in the proximity of the mining zone in the upper basin of the San Jorge River, Colombia. Muñoz and Costa (2012) pointed out that the route of exposure of Ni in humans depends on the chemical form in which it is found, while some soluble species can be expelled quickly through the urine, others can initiate a process of endocytosis of the metal in the cell managing to permeate the membrane and initiating carcinogenic processes. Besides,

J. Marrugo-Negrete · Y. Simanca-Lozano · S. Marrugo-Madrid
Departamento de Química, Facultad de Ciencias Básicas, Grupo de Agua, Química Aplicada y Ambiental, Universidad de Córdoba, Montería, Córdoba, Colombia

S. B. Sujitha (✉)
Escuela Superior de Ingeniería y Arquitectura (ESIA), Unidad Ticoman, Instituto Politécnico Nacional (IPN), Ciudad de México (CDMX), México
e-mail: sbsujitha@yahoo.com

M. P. Jonathan
Centro Interdisciplinario de Investigaciones y Estudios Sobre Medio Ambiente y Desarrollo (CIIEMAD), Instituto Politécnico Nacional (IPN), Ciudad de México (CDMX), México

J. Marrugo-Negrete et al. (eds.), *Remediation and Carbon Capture in Mining Sector*, https://doi.org/10.1007/978-3-031-87883-1_3

Forti et al. (2011) claim that the main route of human exposure to Ni is through inhalation and consumption of contaminated food/water, while exposure through skin contact with soil or water with high concentrations of Ni also proves to be probable routes. Henceforth, human risk assessment of Ni is of foremost importance because of its possible carcinogenic, genotoxic, immunotoxin, and teratogenic effects (Kubrak et al. 2013).

Generally, seafood consumption is one of the prominent exposure routes of heavy metals to humans and nearly 90% of animal protein consumed by coastal and riverine populations comes from fish as a result of low access to other sources (Vargas-Licona et al. 2019). Therefore, the aim of this study was to characterise Ni accumulation in different fish species from the San Jose River Basin (SJR) and Ayapel marshland (AM) and to evaluate the possible human health risks associated due to its consumption.

2 Materials and Methods

2.1 Study Area

Located in the Northwestern Colombia, the San Jorge River Basin (Fig. 1) occupies an area of ~1752 hectares and extends for almost 368 km longitudinally covering 33 municipalities in four adjoining departments. It presents an unimodal rainfall regime, with one dry and one rainy season per year (CVS 2014). The river flows with a minimum and maximum velocity of 24 m^3/s and 697 m^3/s respectively. The annual average precipitation is 2,260 mm, with an average temperature of 27 °C. The river basin also includes several swamps and marshlands, in which the Ayapel Marsh is one of the most important wetland complexes with an area of 145,513 hectares in the middle basin of the SJR and is categorised as a RAMSAR site (CVS 2009). The San Jorge River Basin comprises extensive upstream nickel mining and significant gold mining in the upper San Pedro River (Gracia et al. 2010). On its extensive journey, the river provides economic sustenance to local fishermen, farmers, and loggers from over 70 communities. However, the river basin is deteriorating due to deforestation, erosion, contamination, mining, dumping of garbage, wastewater, and dead animals.

2.2 Collection and Treatment of Samples

Nearly 165 fish samples (no. of species = 11) were acquired from the local fishermen and collection centres located along the SJR. They were specifically collected from the municipalities of San José de Uré, Montelíbano, La Apartada, and

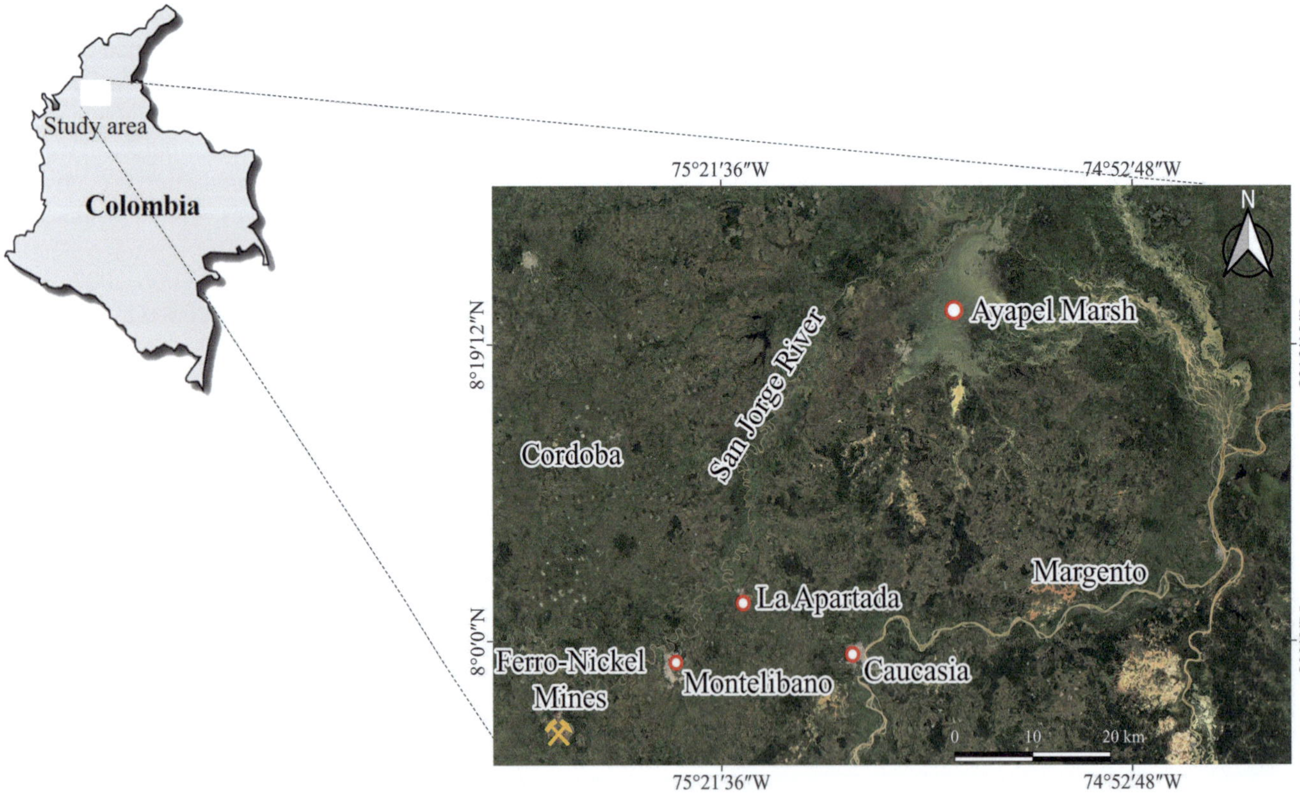

Fig. 1 Map of the study area

Table 1 Mean values ± SD, maximum and minimum concentrations of Ni in the fish species most consumed by the study participants, inhabitants of the banks of the San Jorge River and the Ayapel Marsh, Cordoba-Colombia

Species	San Jorge River							
	Type[1]	Ni (ng/kg w.wt.)		LT (cm)		WT (kg)		n
		M ± SD	Range	M ± SD	Range	M ± SD	Range	
Pseudoplatystoma fasciatum	C	60.3 ± 26.4	104.7–37.2	55.4 ± 8.7	69.5–47.0	1.1 ± 0.6 [2]	2.2–0.7	5
Pimelodus clarias	NC	49.0 ± 32.4	116.3–22.7	21.3 ± 2.5	25.3–14.8	77.9 ± 19.6	0.1–0.04	7
Sorubim cuspicaudus	C	89.0 ± 41.5	124.5–31.7	49.9 ± 10.7	61.1–35.0	842.6 ± 530.9	1.4–0.2	4
Prochilodus magdalenae	NC	63.0 ± 49.3	346.4–10.0	28.3 ± 3.2	41.0–20.7	263.2 ± 89.7	0.5–0.08	18
Leporinus muyscorum	NC	76.3 ± 53.0	212.4–20.6	26.3 ± 2.3	31.1–19.3	181.6 ± 50.0	0.3–0.08	14
Triportheus magdalenae	C	78.1 ± 69.1	218.2–23.3	20.2 ± 0.7	21.7–19.2	63.1 ± 13.5	0.08–0.02	7
Cynopotamus atratoensis	C	73.8 ± 161.2	1064.3–10	24.3 ± 2.4	32.4–21.0	159.1 ± 64.2	0.3–0.09	44
Rhamdia quelen	NC	117.9 ± 99.1	257.9–43.9	26.3 ± 1.8	27.8–23.8	204.5 ± 52.2	0.2–0.1	4
			Ayapel Marsh					
Pimelodus clarias	NC	265.6 ± 161.0	583.7–98.8	19.5 ± 1.8	23.0–16.5	41.3 ± 13.6	0.08–0.02	8
Sorubim cuspicaudus	C	446.3 ± 359.7	1218.0–97.7	48.2 ± 9.0	62.1–31.0	687.2 ± 395.2	1.4–0.1	8
Prochilodus magdalenae	NC	83.9 ± 54.5	128.7–33.4	26.5 ± 5.4	37.7–19.5	197.0 ± 123.3	0.5–0.06	12
Leporinus muyscorum	NC	347.6 ± 136.7	601.9–124.7	28.4 ± 2.7	32.6–22.0	191.4 ± 44.5	0.2–0.09	10
Hoplias malabaricus	C	240.6 ± 96.9	428.5–102.7	27.8 ± 3.0	33.8–24.9	221.1 ± 77.5	0.4–0.1	9
Caquetaia kraussi	C	305.0 ± 117.0	501.6–128.5	18.8 ± 2.8	25.9–16.1	114.3 ± 57.8	0.3–0.07	10
Cyphocharax magdalenae	NC	88.0 ± 26.6	123.1–57.3	13.7 ± 1.4	16.0–12.5	24.4 7.1	0.03–0.02	5

[1]Type: C = Carnivore; NC = Non-Carnivore *N:* Number of specimens

the Ayapel Marsh between April 2017 and March 2018. The samples were individually packed, labelled in plastic bags, and refrigerated at 4 ºC till they reached the laboratory. The fish samples were taxonomically identified in the University de Cordoba (University of Cordoba), Colombia, and preliminary characteristics were recorded with the aid of Fish Base (www.fishbase.org). The collected fish species (Table 1) comprised different trophic levels such as carnivores, omnivores, and iliophages, considering the fact that they are widely consumed in the studied population. Biometric (total weight and length) characteristics of each individual fish sample were measured and noted prior to the analysis of Ni content. Later, the dorsal muscle sample (3 cm and 20 g) was extracted and stored at −20 ºC (Ruíz-Guzmán et al. 2014) until further analysis in the Laboratory of Toxicology and Environmental Management of the University of Córdoba, Colombia.

2.3 Determination of Ni

1 g of muscle tissue sample was lyophilised and was subjected to calcination for 15 h at a temperature of 450 °C. Once calcined 1 mL of concentrated HCl was added and when the acid evaporated, the residue was again calcined for 2 h at 450 °C and later the residue was dissolved with HNO_3 and diluted with distilled water to 25 mL in a volumetric flask (AOAC 2002) for further analysis. Ni concentrations were measured using a graphite furnace Atomic Absorption Spectrometry (GFAAS: *Thermo Scientific™ iCE™ 3500*), following the method SM-3113B (APHA 1995). The accuracy of the method was determined by using DORM-4 (Fish protein certified reference material for trace metals) from the National Research Council of Canada (NRC-CNRC) and the recovery percentage was 103.7 ± 2.1%. The detection limit was calculated as 3 × standard deviation of the targets (n = 3) and the observed result was 20 μg/kg.

2.4 Human Health Risk Assessment

A cross-sectional exploratory study was conducted using the Codex Alimentarius food safety risk assessment (UNEP/WHO 2008). For this purpose, 101 inhabitants (64 men and 37 women) of the San Jorge River Basin in the municipalities of La Apartada, Montelíbano, and San José de Uré, as well as inhabitants of the Ayapel Marsh were surveyed for assessing their fish consumption patterns (species, quantity, and frequency of weekly consumption). Additionally, individual characteristics namely body weight, age, and sex were also documented.

The health risks associated due to the consumption of fish were assessed by calculating the Estimated Daily Intake (EDI) of Ni per kg body weight of the exposed person (μg/kg bw/day) (UNEP 2010; Zamora-Arellano et al. 2017).

$$\mathrm{EDI} = \frac{\mathrm{CNi} \times \mathrm{ID}}{\mathrm{BW}}$$

where CNi corresponds to the average concentration of Ni (μg/kg) in the muscle of the fish analysed and consumed by the participants. DI is the daily fish intake (kg/day) reported for the surveyed population. BW corresponds to the average body weight of the respondents.

The hazard quotient (HQ) is used to estimate the hazard associated with the consumption of food contaminated with chemicals such as heavy metals (USEPA 2000; JECFA 2015) quantitatively. HQ is calculated using the formula.

$$\mathrm{HQ} = \frac{\mathrm{EDI}}{\mathrm{RfD}}$$

where RfD is the Reference Dose of Ni (μg/kg bw/day) in this study a reference of 12 μg/kg bw/day was used (Olivares-Arias et al. 2015; Sadeghi et al. 2019; WHO 2008, 2011; Varol et al. 2017). HQ values >1 indicate potential health risks to humans.

The population's target cancer risk (TR) was also estimated through the probability of developing cancer from exposure to a carcinogen (Asante-Duah 2002); Sadeghi et al. 2019), even though dietary exposure to Nickel does not result in cancerous cells in humans.

$$\mathrm{TR} = \frac{\mathrm{CNi} \times \mathrm{ID} \times 10^{-3} \times \mathrm{CSF} \times \mathrm{EF} \times \mathrm{ED}}{\mathrm{BW} \times \mathrm{ATc}}$$

where CNi is the concentration of the contaminant (μg/g); ID is the daily intake of the contaminated food, in this case, the daily intake calculated for each species in the two evaluated areas; CSF is the oral carcinogenic slope factor of the Integrated Risk Information System (Ni: 1.7 mg/kg/day) (USEPA 2010); EF is the exposure frequency (365 days/year); ED is the exposure duration, that is, the life expectancy of a person, which according to projections until 2020 for Colombians is 74 years (DANE 2007) from which the first two years of life were subtracted; BW is the weight of the inhabitants, and ATc is the average time of exposure to the carcinogen (365 days × 72 years).

Furthermore, the maximum allowable amount of fish (MAF) that could be consumed on a daily basis was also estimated taking into account the high frequency of consumption.

$$\mathrm{MAF} = \frac{\mathrm{RfD} \times \mathrm{ID}}{\mathrm{EDI}}$$

Degree of Contamination

The study also determined the contamination index proposed by Zhang et al. (2019) to gauge the degree of contamination by Ni in the widely consumed fishes of the region using the formula,

$$P_{\mathrm{i}} = \frac{C_i}{S_i}$$

where P_{i} is the pollution index; C_{i} is the concentration of the metal in the fish muscle and S_{i} is the value of the assessment criterion or reference limit (Liang et al. 2011; Zhang et al. 2019). The resultant P_i values are categorised as safe ($P_{\mathrm{i}} \leq 1$); low pollution ($1 < P_{\mathrm{i}} \leq 2$); moderate pollution ($2 < P_{\mathrm{i}} \leq 3$) and high pollution ($P_{\mathrm{i}} > 3$).

2.5 Statistical Analysis

Descriptive statistical analysis was conducted and the obtained results are presented in Table 1. Normality tests (Shapiro–Wilk and Kolmogorov–Smirnov) were carried out and when the assumptions were not fulfilled, Kruskal–Wallis and Dunn tests were performed. To measure the degree of relationship between variables, a Rho Spearman correlation analysis was used. The statistical analysis was developed with IBM SPSS Statistics (version 23) with a significance level of $p < 0.05$.

2.6 Ethical Considerations

The objectives of the present study were submitted to the Ethics Committee of the Faculty of Health Sciences of the University of Cordoba for approval. The process of inclusion and voluntary participation of the individuals were carried out following the national regulations (Resolution No. 8430 of October 4, 1993, Ministry of Health, Republic of Colombia) and international regulations (Declaration of Helsinki and its amendments, World Medical Association

(WMA), Edinburgh, Scotland, October 2000). Signatures and formal consents of all the participants were accomplished before the start of the study.

3 Results

3.1 Ni Concentrations in Fish Muscle Tissue

It is important to consider that this is the first study that presents bioaccumulation of Ni in the muscular tissues of widely consumed fish species of the San Jorge River Basin. The mean ± SD values of Ni in the studied fish species collected from San Jorge River and Ayapel Marsh are presented in Table 1, based on the consumption priorities of the participants. The results of the present study were compared with the findings of Olivares-Arias et al. (2015), in which Ni concentrations in fishes of unpolluted waters varied between 0.4–4 μg/kg w.wt. for confirming whether the Ni values determined in the studied species of this study are within acceptable range. Fishes of the SJR presented Ni concentrations in the order of: *R. quelen* > *S. cuspicaudus* > *T. magdalenae* > *L. muyscorum* > *C. atratoensis* > *P. magdalenae* > *P. fasciatum* > *P. clarias;* with significant difference ($p < 0.05$) among the first species and the others. Highest Ni concentrations were observed in the fishes *S. cuspicaudus* (446.3 ± 359.7 μg/kg w.wt.) and *P. magdalenae* (83.9 ± 54 μg/kg w.wt.) collected from the AM. However, the average Ni concentrations in fishes of the marsh region presented an order of *S. cuspicaudus* > *L. muyscorum* > *C. kraussi* > *P. clarias* > *H. malabaricus* > *C. magdalenae* > *P. magdalenae* (Fig. 2). Statistical results between Ni concentrations, Total length (LT), and weight (WT) are presented in Table 2. In the SJR, the species *S. cuspicaudus* and *R. quelen* showed negative although not significant correlations, similar results were obtained for *C. kraussi* from the AM. Only *H. malabaricus* species caught in the Ayapel Marsh reported positive correlation at a highly significant level ($p = 0.01$).

El-Moselhy et al. (2014) emphasised that the feeding habit of a species is one of the most important factor to be considered for the bioaccumulation of contaminant in an organism. Henceforth, distribution pattern of Ni concentrations was also evaluated based on their feeding habits for better inference (Fig. 3). In the fishes of SJR, no statistically significant difference ($p > 0.05$) was found between the Ni concentrations of species with different feeding habits. However, mean values were higher for fish with non-carnivorous habits (74.2 ± 140.0 μg/kg w.wt.) than for carnivores (70.6 ± 66.4 μg/kg w.wt.) species. On the contrary, in the AM, significant difference ($p < 0.05$) was observed in the Ni levels, as higher values were recorded in the carnivores (325.4 ± 222.7 μg/kg w.wt.) species than in the non-Carnivores (201.4 ± 159.2 μg/kg w.wt.) fishes. Among the four species, common to both the study areas (*P. clarias, S. cuspicaudus, P. magdalenae* y *L. muyscorum*) statistically significant differences were found ($p < 0.05$), the largest being those found in fishes of the marsh. According to the results of the survey carried out in the inhabitants of the areas, 54.5% of the species they consume are classified as carnivorous.

4 Estimation of Potential Risk to Human Health

Survey results showed that the species *P. fasciatum* is highly consumed (360 g/day) by the local communities of San Jorge River Basin, whereas the species *C. kraussi* is highly consumed (385 g/day) by the inhabitants of the Ayapel Marsh. However, among the four species consumed in common by the inhabitants of both the areas, the intake of *L. muyscorum* and *P. clarias* is relatively high (Table 3).

The maximum EDI values in the SJR were reported by the species *S. cuspicaudus* (0.4 μg Ni/kg bw/day), while for Ayapel it was reported by the species *L. muyscorum* (1.8 μg Ni/kg bw/day). However, both values were less than the reference value of 12 μg Ni/kg bw/day. The EDI values calculated for each area indicate that, so far, there is no potential threat or risk to the population consuming any of the eleven species analysed in this study. However, a significant difference ($p < 0.05$) between some species urges the need to monitor their consumption, regarding quantity and frequency, as well as the concentration of Ni in the muscular tissues. The results of the Hazard Quotient were also found to be below the reference value (HQ < 1), although the highest values were recorded for the species *L. muyscorum*. Contamination index (P_i) values for all the fish species presented a strong degree of contamination ($Pi > 3$), where the species *S. cuspicaudus* with a value of $P_i = 37$ is 12-fold higher than the highest degree of contamination.

One of the main reasons for assessing the health risks of those exposed to Ni concentrations is the classification of this metal as a group 1 carcinogen with the capacity to produce genetic abnormalities such as DNA strand breaks (Ahmad and Ashraf 2011; Clancy and Costa 2012; Kas et al. 2008). TR (Target cancer risk) values calculated for the inhabitants of the study area are presented in Suppl. Table T1. The results indicate that the inhabitants of the banks of the SJR and the AM present high risks of suffering one or the other type of cancer due to the consumption of fish with high Ni levels. More specifically, the risk presents greater values with the consumption of species like *L. muyscorum*, *S. cuspicaudus*, and *C. kraussi*.

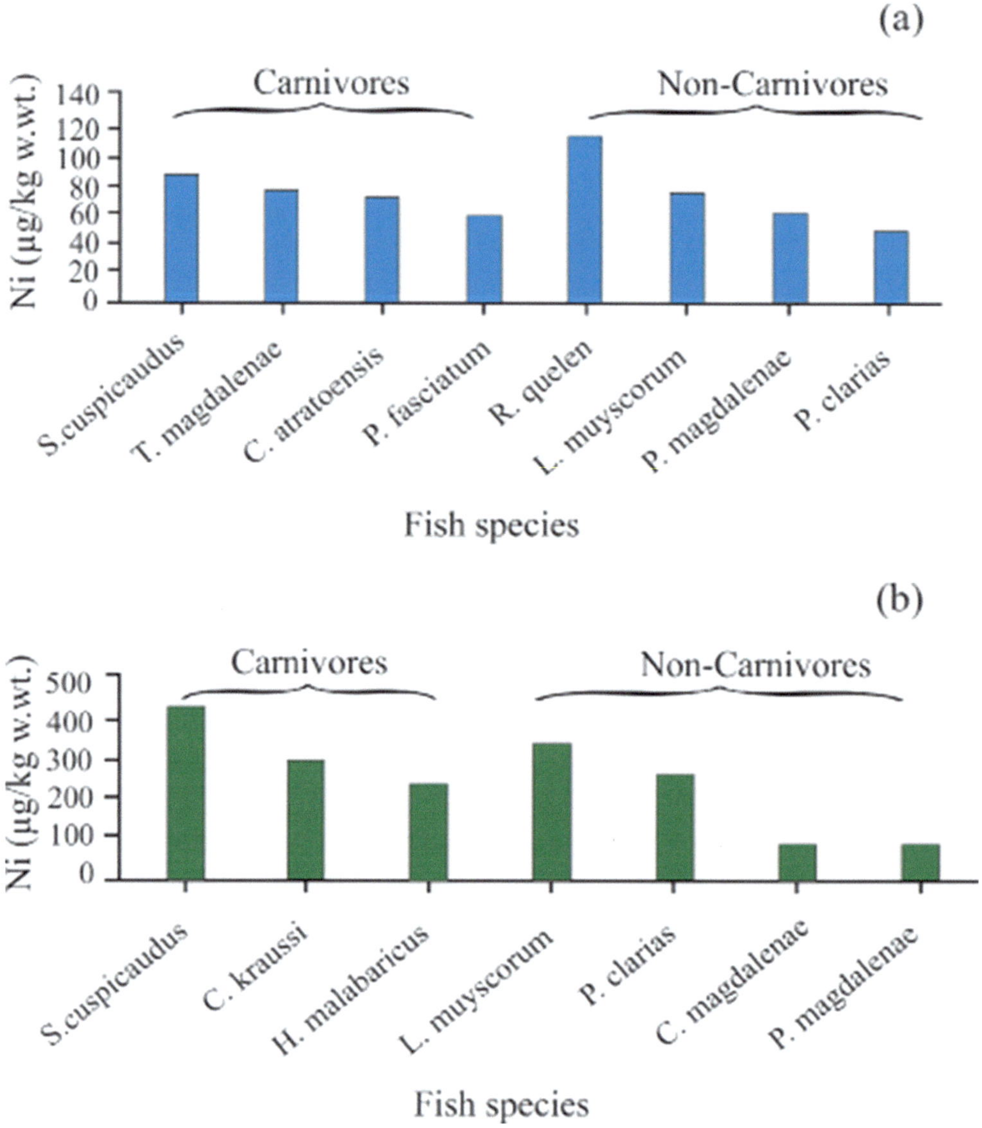

Fig. 2 Ni concentrations (µg/kg w.wt.) in fish caught along the San Jorge River **a** and in the Ayapel Marsh **b**, Córdoba-Colombia

5 Discussion

5.1 Ni Concentrations in Fish Muscle Tissue

Ni concentrations in *P. magdalenae* (220 ± 40 µg/kg w.wt.) from Atrato River Basin, Chocó-Colombia (Palacios-Torres et al. 2020) were higher than the values reported in the present study (San Jorge River: 63.0 ± 49.3 and Ayapel Marsh: 83.9 ± 54 µg/kg w. wt.) probably due to the high bioavailability of Ni in the Atrato River Basin and the larger size of the species analysed (i.e. 32.7 ± 1.2 cm; 522.2 ± 73.9 g). Ni concentrations in the muscular tissues of the studied fish species did not present a clear trophic pattern (Fig. 2), as there was an alternation in the number of species based on the feeding habits from both the study areas. It should also be noted that the Ni levels in *S. cuspicaudus* from the Ayapel Marsh were 3.8 times higher than the levels in *R. quelen* from the river. The absence of biomagnification processes is clearly evidenced by the reduction of Ni concentrations in species of higher tropic levels (Zhang et al. 2013; Fuentes-Gandara et al. 2018).

The Ayapel Marsh is a lentic water body located in the lower middle zone of the San Jorge River, from which it receives water and sediment deposits. Marrugo-Negrete et al. (2014) studied the atmospheric deposition of heavy metals in the mining area of the San Jorge River Basin by recording the precipitation and mass deposition concentrations of Ni in sites adjacent to the extraction of coal and ferronickel. The authors concluded that the sources of Ni contamination are possibly from the presence of metal smelting plants, vehicular traffic that facilitates dispersion, and the soils of the region that are characterised with the presence of Ni laterites. The study also confirmed that the phenomenon of soil degradation, mining activities, run-off, and low pH result in high bioavailability of Ni in lentic

Table 2 Rho Spearman correlation coefficient among Length (r-LT), Weight (r-WT) and Ni concentration contained in fish samples

Species	San Jorge River					Ayapel Marsh				
	r-LT	P	r-WT	P	n	r-LT	P	r-WT	P	n
Pseudoplatystoma fasciatum	0.70	0.19	0.30	0.62	5	NR	NR	NR	NR	NR
Pimelodus clarias	0.32	0.48	0.39	0.38	7	0.12	0.78	0.19	0.65	8
Sorubim cuspicaudus	−0.40	0.60	0.20	0.80	4	0.64	0.09	0.62	0.10	8
Prochilodus magdalenae	0.24	0.33	0.14	0.57	18	0.51	0.09	0.53	0.07	12
Leporinus muyscorum	0.42	0.13	0.09	0.75	14	0.36	0.31	0.41	0.24	10
Triportheus magdalenae	0.31	0.50	0.04	0.94	7	NR	NR	NR	NR	NR
Cynopotamus atratoensis	0.19	0.20	0.19	0.20	44	NR	NR	NR	NR	NR
Hoplias malabaricus	NR	NR	NR	NR	NR	0.80[1]	0.01	0.63	0.07	9
Caquetaia kraussi	NR	NR	NR	NR	NR	−0.08	0.83	0.01	0.99	10
Rhamdia quelen	−0.20	0.80	0.40	0.60	4	NR	NR	NR	NR	NR
Cyphocharax magdalenae	NR	NR	NR	NR	NR	0.56	0.32	0.72	0.17	5

NR: Not Reported in that area; [1]Significant level 0.01

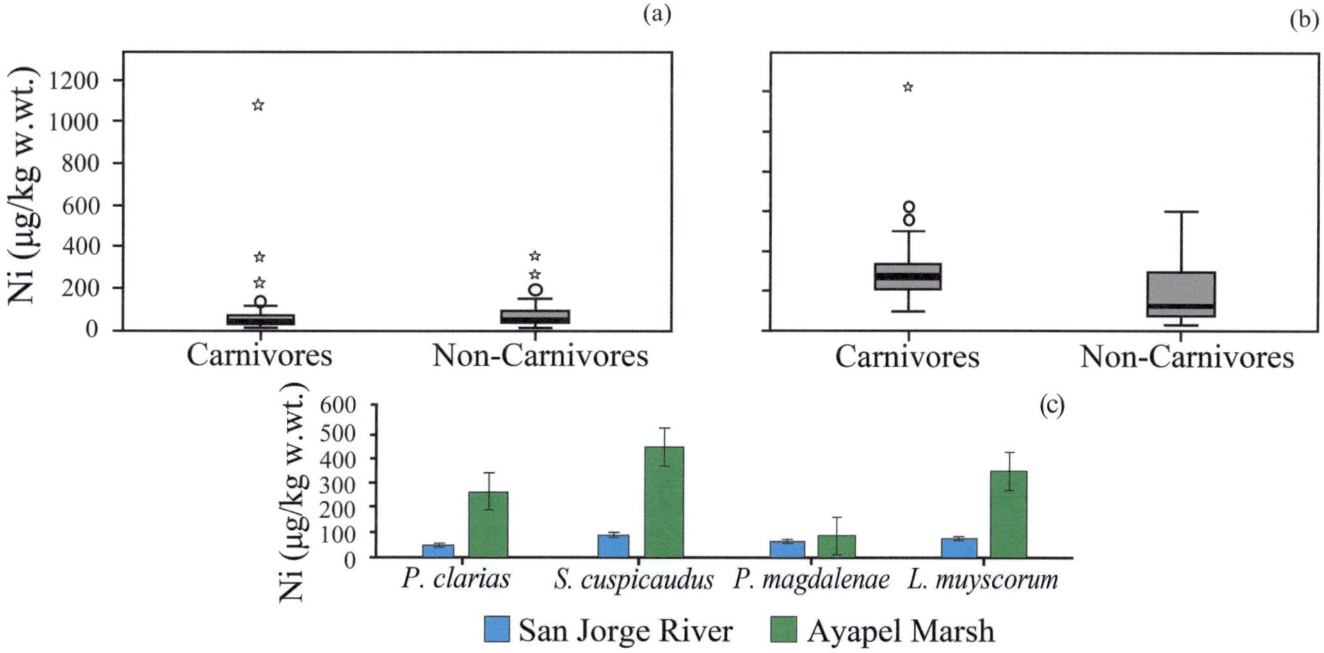

Fig. 3 Distribution of Ni concentration (µg/kg w.wt.) among species with different feeding habits in the San Jorge River bank **a** and in the Ayapel Marsh **b**; **c** Ni concentration among the four species that coincide in the two study areas. Different letters in the columns of the same species indicate significant difference ($p < 0.05$)

water bodies. Taking into consideration the results of the above-mentioned study, the highest concentrations of Ni in the marsh are inferred to the deposition of metals in this site, regardless of whether their origin is related to the nature of the soils, the transport through the river, or anthropic actions. Although no sediment samples were considered for the present study, elevated Ni concentrations in the muscular tissues of fishes at both the study sites are recorded in the samples with the highest trophic relationship to the sediment, such as *S. cuspicaudus* (demersal) and *R. quelen* (benthic).

Positive but not statistically significant correlations found between Ni concentrations and morphometric variables are explained by Sampaio et al. (2005) and Marrugo-Negrete et al. (2008), who agree that the concentrations of some metals tend to be higher in large fish than in small or

Table 3 Potential risk estimation: Daily Intake (DI) (g/day), Estimated Daily Intake (EDI)(ligNi/kg bw/day), Hazard Quotient (HQ), Maximum Allowable fish intake (MAF) (kg) of the most consumed species among the inhabitants of the San Jorge River and the Ayapel Marsh region and the calculated Pollution index (Pi) values (bold values indicate pollution)

Species	DI (g/day)		EDI (jig Ni/kg bw/day)		HQ		MAF(kg)		P_i	
	SJR	AM	SJR	AM	SJR	AM	SJR	AM	SJR	AM
Pseudoplatystoma fasciatum	0.360	–	0.3	–	0.02	–	14	–	5	
Pimelodus clarias	0.179	0.366	0.1	1.4	0.01	0.11	18	3	4	22
Sorubim cuspicaudus	0.302	0.274	0.4	1.7	0.03	0.14	10	2	7	37
Prochilodus magdalenae	0.271	0.339	0.2	0.4	0.02	0.03	14	10	5	7
Leporinus muyscorum	0.183	0.367	0.2	1.8	0.02	0.15	11	2	6	29
Triportheus magdalenae	0.125	–	0.1	–	0.01	–	11	–	7	–
Cynopotamus atratoensis	0.140	–	0.1	–	0.01	–	12	–	6	–
Hoplias malabaricus	–	0.297	–	1.0	–	0.08	–	4	–	20
Caquetaia kraussi	–	0.385	–	1.7	–	0.14	–	3	–	25
Rhamdia quelen	0.174	–	0.3	–	0.02	–	7	–	10	–
Cyphocharax magdalenae	–	0.277	–	0.3	–	0.03	–	10	–	7

SRJ: San Jorge River; AM: Ayapel Marsh. * HQ > 1 represent a potential health risk. Reference limit: 12 μg Ni/kg bw/day

Supplementary Table T1 Estimation of the target cancer risk (TR) of the inhabitants of the areas evaluated by the consumption of each one of the species. TR < 10^{-6} (Insignificant); TR between 10^{-4} y 10^{-6} (Acceptable); TR > 10^{-4} (Unacceptable) (Sadeghi et al. 2019; USEPA 2010)

Species	Target cancer risk										
	1	2	3	4	5	6	7	8	9	10	11
San Jorge River	0.03	0.01	0.03	0.02	0.02	0.01	0.01	–	–	0.02	–
Ayapel Marsh	–	0.11	0.14	0.03	0.15	–	–	0.08	0.14	–	0.03

1: P. fasciatum; 2: P. clarias; 3: S. cuspicaudus; 4: P. magdalenae; 5: L. muyscorum; 6: T. magdalenae; 7: C. atratoensis; 8: H. malabaricus; 9: C. kraussi; 10: R. quelen; 11: C. magdalenae

young fish, due to long exposure time. Palacios-Torres et al. (2020), reported negative correlations between the fish trophic level and the concentrations of several metals, including Ni as the bioaccumulative processes do not depend on the food chain but on the physiological characteristics of each species. This behaviour coincides with the description made by the Agency for Toxic Substances and Disease Registry (ATSDR 2005), which states that for Ni no characteristics have been confirmed that aid in bioaccumulative processes in the muscle of fish or small organisms living in sites contaminated with the metal. However, Carrasco et al. (2011) pointed out that bioaccumulation processes do take place, and are related to the presence of the contaminant in the sediments that serve as food for the initial links in the fish food chain, that is, the detritivorous species that could later be consumed by carnivorous species. Likewise, Shakeri et al. (2019) found positive correlations in the Mediterranean waters between the concentrations of Ni in the fish and the pollution status of the region where they live, which was considered as an indicator of metal bioaccumulation in fish. Fuentes-Gandara et al. (2018) reported significant correlations between Cr, Ni, and Cd concentrations in Colombian Caribbean estuarine ecosystems, with factors such as total length and weight indicating bioaccumulation processes of the metals in the studied fish species.

The distribution pattern of Ni in fish species with different feeding habits in both the zones (Fig. 3) indicates the existence of Ni bioaccumulation in the fish muscle, although a model of metal biomagnification in the species' food chain is not evident. It also confirms the role of *Storage* or *Deposit* of concentrations of heavy metals including Ni, that is evident in the fish species of the Ayapel Marsh. Comparison of Ni concentrations in freshwater systems worldwide (Table 4) clearly depicted the local anthropogenic and geogenic influences. It is important to note that the results of the present study were converted to mg/kg for better inference with other studies and permissible limit values. The results were relatively similar to that of the present study, except in River Kabul, Pakistan that presented the highest Ni values of 112.2 mg/kg d.wt. (Yousafzai et al. 2010). Likewise, the resultant values were also less than the permissible limit of Ni in fish tissues.

Table 4 Comparison of Ni concentrations in muscle tissues of fishes reported worldwide

Place	Country	N	Ni (mg/kg)	Reference
Storm water Ponds, Orlando	USA	15	5.32	Campbell (1994)*
Sakarya basin	Turkey	32	4.40	Barlas (1999)
River Kabul	Pakistan	25	112.2	Yousafzai et al. (2010)
East Kolkata wetlands	India	32	3.03	Bhupande and Mukherjee (2011)
Kelantan River	Malaysia	78	0.01	Hashim et al. (2014)
San Roque Lake	Argentina	37	0.62	Ozparlak et al. (2012)
Plitvice Lake	Croatia	3	ND	Vukosav et al. (2014)*
Indus River	Pakistan	125	0.62	Nawab et al. (2018)
Xiang River	China	98	0.13	Jia et al. (2018)
Agusan River Permissible limits	Philippines	9	4.17	Cabuga et al. (2020)
WHO, 1985			0.6	
EU, 2001			1.5	
FEPA, 2003			0.5	
San Jorge River	Colombia	165	0.07	Present study*
Ayapel Marsh			0.26	

N = number of samples analyzed; All the results reported as dry weight except* in wet basis

5.2 Assessment of Potential Risk to Human Health

The potential risks to human health due to the consumption of contaminated fishes from the SJR were calculated using the maximum allowable level of 12 μg/kg/bw/day of Ni (WHO 2008, 2011). In the present study, the inhabitants of the banks of the San Jorge River and the Ayapel Marsh consumed high quantities of fishes (Table 3). When assessing the potential health risks associated with the consumption of foods contaminated with heavy metals, it is also essential to consider the frequencies at which the local population consumes the contaminated food. Health Canada (2019) indicated that an individual with a high frequency of consumption is the one who ingests the same source of contamination three or more times per week. Likewise, in this study, the inhabitants of both the areas reported a consumption frequency of 2 ± 1 days/week.

The average ingestion rate by the participants involved in the present study was observed to be higher than the values reported by Ruíz-Guzmán et al. (2014) for the population near the Urrá hydroelectric plant in the department of Córdoba, Northern Colombia (178 g/day). Marrugo-Negrete et al. (2020a, b) estimated that the daily fish consumption in the population of the Mojana region was less than 200 g, although their consumption frequencies were 3 ± 1 day/week. Unlike human health risk assessments for the consumption of metals such as mercury and lead, studies on Ni are often difficult to conduct because there is no recommended intake level for this metal because dietary exposure is not considered to be the most critical exposure route (Buxton et al. 2019). However, various studies and food safety agencies have reported a Reference dose value (RfD) for Ni such as 2×10^{-2} μg/g/day (Sadeghi et al. 2019), 20 μg/kg/day (Palacios-Torres et al. 2020; Fuentes-Gandara et al. 2018; USEPA 2016) and 5 μg/kg/day (WHO 1996; Fernández et al. 2007). In addition, the European Committee published a practical guide on metals and alloys used in materials and articles in contact with food, in which it established a specific limit for Ni as 0.14 mg/kg in food (EDQM 2013). Skibniewska et al. (2009) reported that in the Czech Republic, the permissible limit for Ni was considered to be 500 μg/kg, while in Poland there are no recommended maximum daily intake levels. Anke and Seifert (2005) reported that in the German population, the daily intake limit of Ni was established to be 30–400 μg due to the fact that humans generally require ~5–25 μg/day.

Von Stackelberg et al. (2017) pointed out that fish consumption often depends on gender, schooling, purchasing power, time of year, geographical location and, although it seems obvious, the availability of the resource. Henceforth, all the aforementioned factors were considered in the present study to determine the quantities and frequencies of fish consumption by the inhabitants of the study area. Additionally, several participants of the study affirmed that the factors such as climatic conditions (rain and/ or drought), reproductive cycles of the species, and cultural rootedness also influence the consumption, especially it increases the intake of carnivorous species such as *S. cuspicaudus* in specific months of the year (March and April).

Anke et al. (2002) noted that about 20% of women and 1% of men are sensitive to the adverse health effects of Ni. It is important to highlight that the inhabitants of the SJR and the AM may continue to consume fish as confirmed by the estimates of the maximum allowable intake, whereas the evidence of bioaccumulative processes of the species cautions that even if the consumption habits of the inhabitants persist over time, the concentrations of the contaminant may increase posing potential health risks. It should also be noted that the risk assessment is limited only to fish consumption and does not include other food sources of the inhabitants that are likely to contribute to the Ni concentrations that may accumulate in their bodies, thus urging constant monitoring of fish and human populations.

Gusso-Choueri et al. (2018) pointed out that within the management actions that seek to mitigate the systemic risk of suffering the adverse effects of contamination by heavy metals in food, it is fundamental to determine and recognise values such as the number of portions that a community should consume in a given time, thus highlighting that the quality of a population's health is an indicator associated with goals and socio-economic development. Regarding the P_i pollution index, the fact that all species present a high degree of pollution implies the risk of ecological degradation and a decrease in the abundance and biodiversity of the fish community (Authman et al. 2015). Although it has already been stated that there is no risk to human health despite the concentrations presented by the species in muscle, these results raise the obvious urgency of taking action to determine the ecological risk to which the species evaluated are exposed, especially *S. cuspicaudus*. Among the human health effects documented for Ni, and taking into account the results of the estimated target cancer risk (TR) that could be suffered by the inhabitants of the study area (Suppl. Table T1) due to the consumption of fish with high Ni concentrations, there is a warning about its incidence on the reproductive capacity in mammals due to its toxicity and its interference in the neuroendocrine system at the level of the gonads, testicles or ovaries (Olivares-Arias et al. 2015). Kas et al. (2008) described how Ni concentrations increased the rate of miscarriages in exposed women, and the testicular degeneration observed in exposed rats and mice. Several authors claim that Ni could cause multisystemic failure, as well as impaired energy production, skin detoxification, and functional alteration of the colon, kidneys, and liver (Olivares-Arias et al. 2015; Bouchoucha et al. 2013; Das et al. 2010). Researchers also warn the utmost need to investigate Ni toxicity in human intrauterine development since there has been evidence of increased malformations in children with mothers exposed to Ni.

6 Conclusions

Ni concentrations analysed in the fishes of the Ayapel Marsh registered higher values than the fishes of the San Jorge River. Among the studied species, *S. cuspicaudus* presented the highest concentrations of Ni. The present study also revealed the incidences of bioaccumulation processes in the muscular tissues. Potential risk estimates of the inhabitants reveal less hazards allowing the population to continue with the regular consumption of the fish species. Even though dietary exposure is not considered to be a critical route of Ni contamination in humans, the results of the present study are only indicative as Ni toxicity in humans due to seafood consumption is mainly excreted via urine causing no cancerous effects. As all the studied fish species presented a strong degree of Ni contamination due to the regional anthropogenic and natural processes, the authors call for long-term monitoring of the river basin for better conservation of the local biodiversity and human health.

Acknowledgements The authors would like to thank the Universidad de Córdoba, Monteria, Colombia, and The Administrative Department of Science, Technology, and Innovation (Departamento Administrativo de Ciencia, Tecnología e Innovación – Colciencias), Bogotá, Colombia. SBS and MPJ thanks Sistema Nacional de Investigadores, CONACyT, Mexico. MPJ thanks Instituto Politécnico Nacional (EDI and COFAA), México. This article is a contribution (partial) from the Earth System Science Group (ESSG), Mexico & Chennai, India (Participating members: SBS & MPJ).

Ethical Approval We would also like to state that the study does not involve human subjects.

References

Ahmad, M. and Ashraf, M. (2011). Essential roles and hazardous effects of nickel in plants. *Reviews of Environmental Contamination & Toxicology*, **214:** 125–167.

Anke, M., Mueller, M., Truepschuch, A. and Mueller, R. (2002). Intake and effects of cadmium, chromium and nickel in human. J. Commod. Sci., **4:** 41–63.

Anke, M. and Seifert, M. (2005). Nickel – in fuer die Flora, Fauna and den Menschen essentielles und toxisches Spuren element. In: Experimentelle Modelle der Spuren element for schung, (Eds. Windisch, W., Plitzner, C.,) Muenchen: Herber Utz Verlag, pp. 133–78.

AOAC. (2002). Official Method of Analysis. 16th Edition, Association of Official Analytical, Washington DC.

Asante-Duah, K. (2002). Public Health risk assessment for human exposure to chemicals. Environmental Pollution. Edition, Kindle edition. Springer Netherlands, 2002, pp. 352.

ATSDR (Agencia para Sustancias Tóxicas y Registro de Enfermedades) (2005). Resumen de salud pública Níquel, CAS#: 7440–02–0. https://www.atsdr.cdc.gov/es/phs/es_phs15.pdf.

Authman, M., Zaki, M. and Khallaf, E. (2015). Use of Fish as Bio-indicator of the Effects of Heavy Metals Pollution. *Journal of Aquatic Research & Development*, **6(4):** 328–340.

Barlas, N. (1999). A pilot study of heavy metal concentration in various environments and fishes in the upper Sakarya River basin, Turkey. *Environmental Toxicology*, **14:** 367–373.

Bhupande, K. and Mukherjee, D.P. (2011). Assessment of human health risk for Arsenic, Copper, Nickel, Mercury and Zinc in fish collected from Tropical wetlands in India. *Advances in Life Science & Technology*, **2:** 13–24.

Blewett, T.A. and Leonard, E.M. (2017). Mechanisms of nickel toxicity to fish and invertebrates in marine and estuarine waters. *Environmental Pollution*, **223:** 311–322.

Bouchoucha, A., Terbouche, A., Zaouani, M., Derridj, F. and Djebbar, S. (2013). Iron and nickel complexes with heterocyclic ligands: Stability, synthesis, spectral characterization, antimicrobial activity, acute and subacute toxicity. *Journal of Trace Elements & Medicine Biology*, **27:** 191–202.

Buxton, S., Garman, E., Heim, K.E., Lyons-Darden, T., Schlekat, C., Taylor, M.D. and Oller, A.R. (2019). Concise Review of Nickel Human Health Toxicology and Ecotoxicology. *Inorganics*, **7:** 89–127.

Cabuga, Jr. C.C., Capangpangan, R.Y. and Jumawan, J.C. (2020). Levels of heavy metals in fish and sediments from different salinity gradients of Lower Agusan River to Butuan Bay, Caraga, Philippines. Environment Asia, **13(2):** 88–100.

Campbell, K.R. (1994). Concentrations of heavy metals associated with urban runoff in fish living in storm water treatment ponds. *Archives of Environmental Contamination & Toxicology*, **27:** 352–356.

Carrasco, L., Benejam, L., Bayona, J. and Diez, S. (2011). Methylmercury levels and bioaccumulation in the aquatic food web of a highly mercury-contaminated reservoir. *Environmental International*, **37:** 1213–1218.

Chen, C.Y., Wang, Y.F., Lin, Y.H. and Yen, S.F. (2003). Nickel-induced oxidative stress and effect of antioxidants in human lymphocytes. *Archives of Toxicology*, **77:** 123–130.

Clancy, H. and Costa, M. (2012). Nickel: A pervasive carcinogen. *Future Oncology*, **8:** 1507–1509.

Coman, V., Robotin, B. and Ilea, P. (2013). Nickel recovery/removal from Industrial wastes: A review. *Resources Conservation and Recyling*, **73:** 229–238.

CVS (Corporación Autónoma Regional de los Valles del Sinú y del San Jorge) (2014). Plan de acción para la temporada invernal en el departamento de Córdoba - PATI CVS. Montería, Colombia.

CVS (Corporación Autónoma Regional de los Valles del Sinú y del San Jorge) (2009). Environmental assessment of the hydrographic basin of San Jorge River, Monteria, Colombia.

DANE (Departamento Administrativo Nacional de Estadística) (2007). Proyecciones de población 2005–2020., Available online: https://www.dane.gov.co/files/ investigaciones/poblacion/proyepobla06_20/8Tablasvida1985_2020.pdf. (Accessed on 15 08 2019)

Das, K., Das, S. and Dhundasi, S. (2010). Nickel: Molecular diversity, application, essentiality and toxicity in human health. In: Blanc, G., & Moreau, D., (Eds). Biometals: Molecular structures, binding properties and applications, 33–58. New York. NY: Nova Science Publishers

EDQM (Comité Europeo para Materiales y Artículos en Contacto con Alimentos) (2013). Metals and alloys used in food contact materials and articles The Council of Europe Guidelines in a nutshell. Available online: https://www.edqm.eu/en/food-contact-materials, (Accessed on 15 11 2019)

Eisler, R. (1998). Nickel Hazards to Fish, Wildlife, and Invertebrates: a Synoptic Review. US Geological Survey, Biological Resources Division. Biological Science Report, 1998–2001.

El-Moselhy, K., Othman, A., Abd El-Azem, H. and El-Metwally, M. (2014). Bioaccumulation of heavy metals in some tissues of fish in the Red Sea, Egypt. Egyptian *Journal of Basic & Applied Sciences*, **1-2:** 97–105.

EU (European Union) (2001). Commission Regulation as regards heavy metals, Directive, 2001/22/EC, No: 466.

EC (European Commission) (2008). European Union risk assessment report on nickel, nickel sulphate, nickel carbonate, nickel chloride, nickel dinitrate. Final report May 2008. Prepared by Danish Environmental Protection Agency on behalf of the European Union, Copenhagen, Denmark.

EU (European Union) (2019). Acute water borne nickel toxicity in the European Union.

FEPA (Federal Environmental Protection Agency) (2003). Guidelines and standards for environmental pollution control in Nigeria, p. 238.

Fernández, I., Mateos, C., Aguilar, M. and Martínez, M. (2007). Ingesta diaria de níquel entre jóvenes españoles Valoración del riesgo toxicológico. *Revista de Toxicologia.*, **24**: 10–13.

Forti, E., Salovaara, S., Cetin, Y., Bulgheroni, A., Pfaller, R. and Prieto, P., (2011). In vitro evaluation of the toxicity induced by nickel soluble and particulate forms in human airway epithelial cells. *Toxicology in Vitro*, **25:** 454–461.

Fuentes-Gandara, F., Pinedo-Hernández. J., Marrugo-Negrete, J. and Díez, S. (2018). Human health impacts of exposure to metals through extreme consumption of fish from the Colombian Caribbean Sea. *Environmental Geochemistry & Health*, **40(1):** 229–242.

Gracia, H., Marrugo-Negrete, J. and Alvis, E. (2010). Contaminación por mercurio en humanos y peces en el municipio de Ayapel, Córdoba, Colombia. *Revista Facultad Nacional de Salud Pública*, **28(2):** 118–124.

Gusso-Choueri, P., Araújo, G., Cruz, A., Stremel, T., Campos, S., Abessa, D. and Choueri, R. (2018). Metals and arsenic in fish from a Ramsar site under past and present human pressures: Consumption risk factors to the local population. *Science of the Total Environment*, **628–629:** 621–630.

Hashim, R., Song, T.H., Muslim, N.Z.MD. and Yen. T.P. (2014). Determination of heavy metal levels in fishes from the lower reach of the Kelantan River, Kelantan, Malaysia. *Tropical Life Sciences & Research*, **25(2):** 21–39.

Health Canada (2019). Mercury in Fish: Consumption Advice: Making Informed Decisions about Fish. Available online: https://www.canada.ca/en/health-canada/services/food-nutrition/food-safety/chemical-contaminants/environmental contaminants/mercury/mercury-fish.html. (Accessed on 18 01 2020).

INGEOMINAS (Instituto de Investigación e Información Geocientífica Minero-Ambiental) (2005). Environmental assessment of River Basin San Jorge. Chapter 9. Mining Resources, Colombia, 87p.

JECFA (Joint FAO/WHO Expert Committee on Food Additives) (2015). Summary and Conclusions of the meetings of the joint FAO/WHO. Available online: https://apps.who.int/food-additivescontaminants-jecfa-database/Search.aspx. (Accessed on 30 11 2019).

JECFA (Joint FAO/WHO Expert Committee on Food Additives) (2018). Summary and Conclusions of the meetings of the joint FAO/WHO. Available online: https://apps.who.int/food-additivescontaminants-jecfa-database/Search.aspx. (Accessed on 30 11 2019).

Jia, Y., Wang, L., Cao, J., Li, S. and Yang, Z. (2018). Trace elements in four freshwater fish from a mine-impacted river: spatial distribution, species-specific accumulation, and risk assessment. *Environmental Science & Pollution Research*, **25:** 8861–8870.

Kas, K., Das, S. and Dhundasi, S. (2008). Nickel, its adverse health effects and oxidative stress. *Indian Journal of Medical Research*, **128:** 412–425.

Kubrak, O.I., Husak, V.V., Rovenko, B.M., Poigner, H., Kriews, M., Abele, D. and Lushchak, V.I. (2013). Antioxidant system efficiently protects goldfish gills from Ni^{2+} -induced oxidative stress. *Chemosphere*, **90:** 971–976.

Lebrun, J.D., Perret, M., Uher, E., Tusseau-Vuillemin, M-H. and Gourlay-France, C. (2011). Waterborne nickel bioaccumulation in Gammarus pulex: Comparison of mechanistic models and influence of water cationic composition. *Aquatic Toxicology*, **104(3–4):** 161–162.

Leonard, E.M. and Wood, C.M. (2013). Acute toxicity, critical body residues, Michaelis- Menten analysis of bioaccumulation, and ion regulatory disturbance in response to waterborne nickel in invertebrate species: *Chironomus riparius, Lymnaea stagnalis, Lumbriculus variegatus* and *Daphnia pulex. Comparative Biochemistry Physiology*, C **158:** 10–21.

Liang, J., Chen, C., Song, X., Han, Y. and Liang, Z. (2011). Assessment of heavy metal pollution in soil and plants from Dunhua sewage irrigation area. *International Journal of Electrochemical Science*, **6(11):** 5314–5324.

Maceda-Veiga, A., Monroy, M. and De Sostoa, A. (2012). Metal bioaccumulation in the Mediterranean barbel (*Barbus meridionalis*) in a Mediterranean river receiving effluents from urban and industrial wastewater treatment plants. *Ecotoxicology & Environmental Safety*, **76:** 93–101.

Madrid, G., Gracia, L., Marrugo-Negrete, J. and Urango, I. (2011). Genotoxicity studies of heavy metals: Hg, Zn, Cu, Pb y Cd related to mining operations on residents of San Jorge basin, Department of Cordoba, Colombia. *Revista Asociada Col Ciencias*, **23:** 103–111.

Marrugo, J., Lans, E., Doria, G., Bello, L., Castillo, M., Cortes, F. and Pinedo, J. (2006). Impacto ambiental por contaminación con níquel, mercurio y cadmio en aguas, peces y sedimentos en la cuenca del Río San Jorge, en el Departamento de Córdoba. Universidad de Córdoba, Centro de Investigaciones – CIUC. Montería, Colombia, pp. 109.

Marrugo-Negrete, J., Benitez, L., Olivero-Verbel, J., Lans, E. and Gutierrez, F. (2010). Spatial and seasonal mercury distribution in the Ayapel Marsh, Mojana region, Colombia. *International Journal of Environmental Health Research*, **20(6):** 451–459.

Marrugo-Negrete, J., Lans, E. and Benítez. L. (2007). Finding of mercury in fish from the Ayapel marsh, Cordoba, Colombia. *Revista MVZ Córdoba*, **12(1):** 878–886.

Marrugo-Negrete, J., Olivero-Verbel, J., Lans, E. and Benitez, L. (2008). Total mercury and methylmercury concentrations in fish from the Mojana region of Colombia. *Environmental Geochemistry & Health*, **30:** 21–30.

Marrugo-Negrete, J., Urango-Cardenas, I., Burgo-Núñez, S. and Díez, S. (2014). Atmospheric deposition of heavy metals in the mining area of the San Jorge river basin, Colombia. *Air Quality & Atmospheric Health*, **7(4):** 577–588.

Marrugo-Negrete, J., Pinedo-Hernandez, J., Marrugo-Madrid, S. and Díez, S. (2020). Assessment of trace element pollution and ecological risks in a river basin impacted by mining in Colombia. *Environmental Science & Pollution Research*, https://doi.org/10.1007/s11356-020-10356-4.

Marrugo-Negrete, J., Vargas-Licona, S., Ruíz-Guzmán, J., Marrugo-Madrid, S., Bravo, A. and Díez, S. (2020). Human health risk of methylmercury from fish consumption at the largest floodplain in Colombia. *Environmental Research*, **182:** 109050.

Muñoz, A. and Costa, M. (2012). Elucidating the mechanisms of nickel compound uptake: A review of particulate and nano-nickel endocytosis and toxicity. *Toxicology & Applied Pharmacology*, **260(1):** 1–16.

Nakajima, K., Nansai, K., Matsubae, K., Tomita, M., Takayanagi, W. and Nagasaka, T. (2017). Global land use change hidden behind nickel consumption. *Science of the Total Environment*, **586:** 730–737.

Nawab, J., Khan, S. and Xiaoping, W. (2018). Ecological and health risk assessment of potentially toxic elements in the major rivers of Pakistan: general population vs. fishermen. *Chemosphere*, **202:** 154–164.

Olivares-Arias, V., Valverde-Som, L., Quiros-Rodríguez, V., García-Romero, R., Muñoz, N., Navarro-Alarcón, M. and Cabrera-Vique, C. (2015). Nickel in food and influencing factors in its levels, intake, bioavailability and toxicity: A review. *CyTA & Journal Food*, **13(1):** 87–101.

Özparlak, H., Arslan, G. and Arslan, E. (2012). Determination of some metal levels in muscle tissue of nine fish species from Beyşehir Lake, Turkey. *Turkish Journal of Fisheries & Aquatic Sciences*, **12:** 761–770.

Palacios-Torres, Y., De la Rosa, J. and Olivero-Verbel, J. (2020). Trace elements in sediments and fish from Atrato River: An ecosystem with legal rights impacted by gold mining at the Colombian Pacific. *Environmental Pollution*, **256:** 113290.

Palermo, F., Risso, W.E., Simonato, J.D. and Martinez, C.B.R. (2015). Bioaccumulation of nickel and its biochemical and genotoxic effects on juveniles of the neotropical fish *Prochilodus linneatus. Ecotoxicology & Environmental Safety*, **116:** 19–28.

Pane, E.F., Richards, J.G. and Wood, C.M. (2003). Acute waterborne nickel toxicity in the rainbow trout (Oncorhynchus mykiss) occurs by a respiratory rather than ion-oregulatory mechanism. Aquatic Toxicology, **63:** 65–82.

Pérez-Rodríguez, P., De Blas, E., Soto, B., Pontevedra-Pombal, X. and López-Periago, J. (2014). The soil use conflict and food quality. *CyTA & Journal of Food*, **9(4):** 342–350.

Ragsdale, S.W. (1998). Nickel biochemistry. *Current Opinion in Chemical Biology*, **2:** 208–215.

Ruiz-Guzmán, J., Marrugo-Negrete, J. and Díez, S. (2014). Human Exposure to Mercury Through Fish Consumption: Risk Assessment of Riverside Inhabitants of the Urrá *Reservoir, Colombia. Human Ecology Risk Assessment International Journal*, **20(5):** 1151–1163.

Sadeghi, P., Loghmani, M. and Afsa, E. (2019). Trace element concentrations, ecological and health risk assessment in sediment and marine fish *Otolithes ruber* in Oman Sea, Iran. *Marine Pollution Bulletin*, **140:** 248–254.

Sampaio, D., Luccote, M., Roulet, M., Poirier, H., Margler, D., Oliveira, E. and Crossa, M. (2015). Trophic structure and bioaccumulation of mercury in fish of three natural lakes of the Brazilian Amazon. *Water Air Soil Pollution*, **165:** 77–94.

Shakeri, A., Fard, M. and Mehrabi, B. (2019). Arsenic and heavy metals (HMs) occurrence in sediment and fish of a geothermal area; Case study: Khiav River, Ardebil Province - NW of Iran. *Journal of Geochemical Exploration*, **208:** 106347.

Skibniewska, K., Guziur, J., Marzec, Z., Zaręba, S., Grzybowski, M. and Szarek, J. (2009). Nickel in the muscle tissues of freshwater fish from north eastern Poland should not cause human health concerns. Toxicol*ogical & Environmental Chemistry*, **914:** 773–778.

Sun, Z., Gong, C., Ren, J., Zhang, X., Wang, G., Liu, Y., Ren, Y., Zhao, Y., Yu, Q., Wang, Y. and Hou, J. (2020). Toxicity of nickel and cobalt in Japanese Flounder. *Environmental Pollution*, **263:** 114516.

Takeda, K., Marumoto, K., Minamikawa, T., Sakugawa, H. and Fujiwara, K. (2000). Three-year determination of trace metals and the lead isotope ratio in rain and snow depositions collected in Higashi-Hiroshima, Japan. *Atmospheric Environment*, **34:** 4525–4535.

Topal, A., Atamanalp, M., Oruç, E., Halıcı, M.B., Sisecioglu, M., Erol, H.S., Gergit, A. and Yılmaz, B. (2015). Neurotoxic effects of nickel chloride in the rainbow trout brain: assessment of c-Fos activity, antioxidant responses, acetylcholinesterase activity, and histopathological changes. *Fish Physiology & Biochemistry*, **41(3):** 625–634.

UNEP (United Nations Environment Programme) (2010). Guidance for identifying population at risk from mercury exposure, UNEP (DTIE)/ Hg/INC.2/INF/3. UNEP, Geneva, Switzerland.

UNEP/WHO (United Nations Environment Programme/World Health Organization) (2008). Guidance for Identifying Populations at Risk from Mercury Exposure. pp.176.

UPME (Unidad de Planeación Minero Energética) (2018). Níquel - Unidad de Planeación Minero Energética UPME Tableau Public. Available online: https://public.tableau.com/profile/upme#!/vizhome/Niquel/Historia1 (Accessed on 12 12 2019).

USEPA (Environmental Protection Agency) (2000). Guidance for assessing chemical contaminant data for use in fish advisories, Risk assessment and fish consumption limits. EPA 823-B-00–008, 3rd ed. Vol. 2. p. 383. Washington. DC: USEPA Office of Water. Available online: https://www.epa.gov/sites/production/files/2015-06/documents/volume2.pdf. (Accessed on 21 12 2019).

USEPA (Environmental Protection Agency) (2010). Risk-Based Concentration Table. Available online: http://www.epa.gov/reg3hwmd/risk/ human/index.htm (Accessed on 12 12 2019).

USEPA (Environmental Protection Agency) (2016). Integrated risk information system. CRC. Available online: http://www.epa.gov/iris/. (Accessed on 21 01 2020).

USEPA (Environmental Protection Agency) (2019). Guidance for assessing chemical contaminant data for use in fish advisories, Risk assessment and fish consumption limits. EPA 823-B-00–008, 3rd ed. Vol. 2. p. 383. Washington. DC: USEPA Office of Water, 2000. Available online: https://www.epa.gov/sites/production/files/2015-06/documents/volume2.pdf. (Accessed on 21 12 2019).

USEPA (Environmental Protection Agency) (2020). Integrated risk information system. CRC, 2016. Available online: http://www.epa.gov/iris/ (Accessed on 21 01 2020).

Vargas-Licona, S. and Marrugo-Negrete, J. (2019). Mercury, Methylmercury and other heavy metals in fish in Colombia: Risk from ingestion. Acta Biologia Colombia., **24(2):** 232–242.

Varol, M., Kurt, G. and Alp, A. (2017). Heavy metal and arsenic concentrations in rainbow trout (*Oncorhynchus mykiss*) farmed in a dam reservoir on the Firat (Euphrates) River: Risk-based consumption advisories. *Science of the Total Environment*, **599–600:** 1288–1296.

Von Stackelberg, K., Li, M. and Sunderland, E. (2017). Results of a national survey of high-frequency fish consumers in the United States. *Environmental Research*, **158:** 126–136.

Vukosav, P., Mlakar, M., Cukrov, N., Kwokal, Ž., Pižeta, I., Pavlus, N., Špoljarić, I., Vurnek, M., Brozinčević, A. and Omanović, D. (2014). Heavy metal contents in water, sediment and fish in a karst aquatic ecosystem of the Plitvice Lakes National Park (Croatia). *Environmental Science & Pollution Research*, **21:** 3826–3839.

WHO (World Health Organization) (1985). Guidelines for Drinking Water Quality, Recommendation, vol. 1, Geneva, pp. 130.

WHO (World Health Organization) (1996). Trace elements in human nutrition and health. World Health Organization. Geneva.

WHO (World Health Organization) (2008). Guidelines for drinking-water quality incorporating 1st and 2nd addenda, 3rd ed., Vol. 1, Recommendations. Geneva.

WHO (World Health Organization) (2010). Más salud por el dinero, La Financiación de los Sistemas de Salud. El camino hacia la cobertura universal. Informe Sobre La Salud En El Mundo. pp.67–88.

WHO (World Health Organization) (2011). Guidelines for drinking water quality. 4ft ed. Geneva.

Wood, C.M. (2012). An introduction to metals in fish physiology and toxicology: basic principles. *In:* Wood, C.M., Farrell, A.P., Brauner, C.J. (Eds.), Homeostasis and Toxicology of Essential Metals, Fish Physiology, 31A. Elsevier Inc, pp. 253–289.

Yang, Z., Wang, Y., Shen, Z., Niu, J. and Tang, Z. (2009). Distribution and speciation of heavy metals in sediments from the mainstream tributaries and lakes of the Yangtze River catchment of Wuhan, China. *Journal of Hazardous Material*, **166(2–3):** 186–1194.

Yousafzai, A.M., Chivers, D.P., Khan, A.R., Ahmad, I. and Siraj, M. (2010). Comparison of heavy metals burden in two freshwater fishes *Wallago attu* and *Labeo dyocherlius* with regard to their feeding habits in natural ecosystem. *Pakistan Journal of Zoology*, **42(5):** 537–544.

Zamora-Arellano, N., Ruelas-Inzunza, J., García-Hernández, J., Ilizaliturri-Hernández, C. and Betancourt-Lozano, M. (2017). Linking fish consumption patterns and health risk assessment of mercury exposure in a coastal community of NW Mexico. *Human and Ecological Risk Assessment*, **23(6):** 1505–1521.

Zhang, T., Xu, W., Lin, X., Yan, H., Ma, M. and He, Z. (2019). Assessment of heavy metals pollution of soybean grains in North Anhui of China. *Science of the Total Environment*, **646:** 914–922.

Zhang, X., Xu, Q., Man, S., Zeng, X., Yu, Y., Pang, Y., Sheng, G. and Fu, J. (2013). Tissue concentrations, bioaccumulation, and biomagnifications of synthetic musks in freshwater fish from Taihu Lake, China. *Environmental Science & Pollution Research,* **20:** 311–322.

Zuluaga-Rodríguez, J., Gallego-Ríos, S. and Ramírez-Botero, C. (2015). Content of Hg, Cd, Pb and As in fish Species: A review, *VITAE*, **22(2):** 148–159.

Mining Hazards and Human Health Risk: A Case Study of Stone Mining in Hamirpur District Himachal Pradesh, India

Shanta Kumar, Riju, and Anurag Linda

1 Introduction

The earth comprises hydrosphere, lithosphere, atmosphere, and biosphere. These spheres interact with one another and also maintain a delicate balance among themselves (Chaurasia et al. 2009; Central Pollution Control Board [CPCB] 2009; Himachal Pradesh State Pollution Control Board [HPSPCB] 2014). There are different biotic and abiotic components in the environment that form structure and functioning of the environment. With increased refinement in the field of science and technology, there is an over exploration and exploitation of natural resources, resulting in disruption of structure and functioning of the environment. In addition, with an increase in anthropogenic developments, our natural resources are deteriorating day by day, and our planet earth is already paying a hefty price for these progressive unsustainable developments.

On the earth's surface, river has a lead role in the aquatic and terrestrial ecosystem (Kamboj et al. 2018). In recent times, India is facing the biggest problem of air, soil, and water pollution. Especially at the time of the Industrial Revolution (eighteenth century–twentieth century), the condition of air, soil, and water got worst due to least awareness on environmental safety and related issues (HPSPCB 2014). Increased mining and extraction activities of stone, gravel, and sand from riverbeds lead to high pressure on the river ecosystem (Kamboj et al. 2018). Throughout the world, stone mining is executed for the construction of roads, buildings, pavements, and other industrial uses (CPCB 2009; Kamboj et al. 2017; Mehta 2002).

Economic growth and ever increasing demands of resources in the twentieth century have exaggerated the future availability of natural resources (Manzoor and Khan 2020). Several new problems directly related to environmental degradation, human health hazards and loss of biodiversity have evolved (Patil 2001).

India is rich in natural resources mined throughout the country. The Himalayas also hold plenty of natural resources, and their exploitation in recent decades has disturbed the ecology of the fragile Himalayan ecosystem. The impact of stone mining is a good example to evaluate the impact of mining in the Himalayan ecosystem. In India, there are about 12,000 stone-crushing plants, as reported in the year 2001, which has a projected annual gross revenue of more than US$1 billion (CPCB 2009; HPSPCB 2014). The stone-crushing plants have a significant contribution in the industrial sector. They produce raw materials in the form of crushed stones of several sizes on demand, which are further used in a variety of construction operations, including roads, highways, bridges, buildings, and canals (Pal and Mandal 2019a; Talwar et al. 2017). During the operational phase of stone mining, stone dust is produced, a primary source of air-borne pollution. Its intensity is determined by several elements related to the local micro-climatic conditions of the area (Manzoor and Khan 2020). Owing to fugitive emission of air pollutants from stone-crushing plants, acute environmental issues are coming up which are deteriorating human and plant health. For example, there is an increase in respiratory diseases, such as silicosis, cellular breakdown in the lungs, pneumonic tuberculosis (TB), and other aspiratory and non-pneumonic ailments (HPSPCB 2014, 2019). The air-borne pollution is also retarding the growth of vegetations and hampering aesthetic values of an area by degrading the quality of water and soil. These activities have impacted the financial highlights of a nation and regional residents.

S. Kumar · A. Linda (✉)
Department of Environmental Sciences, Central University of Himachal Pradesh, Dharamshala, Himachal Pradesh, India
e-mail: anuraglinda.cuhp@hpcu.ac.in

Riju
Department of Environmental Studies, Panjab University, Chandigarh, India

A. Linda
Department of Environmental Sciences, Central University of Jharkhand Ranchi, Ranchi, Jharkhand, India

J. Marrugo-Negrete et al. (eds.), *Remediation and Carbon Capture in Mining Sector*, https://doi.org/10.1007/978-3-031-87883-1_4

2 Environmental Degradation and Stone Mining

Minerals are unreplenishable and have sparse reserves, but still they are an essential commodity for industries (Mehta 2002). Unsustainable mineral exploitation of the nature is creating ecological imbalance with a hazardous impact on the environment, because mining, wildlife, fisheries, forest, water balance, local climate, and pattern of rainfall are getting affected at different intensities (Karmakar and Das 2012). The mining units in India have been aggravated by a number of factors, such as environmental clearances, health issues, and safety-related problems (Ambastha et al. 2014). Several events of causalities have been reported due to underground and surface mining of coal, stone, and other minerals in last few decades (CPCB 2009; Mehta 2002). Environmental degradation generally influences the flora and fauna of the area, and apparent impact on soil quality has been observed as well in the mining area because of the absence of appropriate preparatory plans and carelessness in mining guidelines follow-up (Dhar and Ahmed 1993).

During stone mining operations, clouds of dust are produced by stone crusher units. Approximately around 10–11 tonnes/day of dust is documented during the mining operation in the middle catchment of the Dwarka River Basin in Eastern India. It is not only harmful for the soil, water, and air but is also hazardous for the human health (Pal and Mandal 2019a). The dust produced during mining can even increase the ambient temperature by 1–1.5 °C and the concentration of particulate matter of 10 microns (PM_{10}) in the lower atmosphere by >180 mg/m^3, which severely affects vegetation health (Haque and Singh 2017; Pal and Mandal 2019a; Singh et al. 2009). According to Pal and Mandal (2019b), level of noise pollution goes beyond 85 db; the authors also reported a loss of 15–60% in agriculture and fish productivity because of degradation in soil and water pollution. The dust produced due to different mining activities settles on plants and changes the rate of photosynthesis, foliar injuries, transpiration, and uptake and accumulation of mineral elements from soil (Lerman and Darley 1975).

According to, CPCB (2009), in India, small-scale stone crushers are mainly labour-dependent and are operated manually. Approximately 12,000 crushers are functioning in India and have a significant role in the local economy. In the near future, their numbers are expected to grow due to different developmental activities. The increasing number of stone crushers can have deleterious effects on air, water, and soil quality because they emit dust particles into the atmosphere that finally get transferred into soil and water. Diseases such as silicosis, bronchitis, and TB have been reported nearby the mining areas (Haque and Singh 2017; Singh et al. 2009).

Health hazards, both physical and physiological, can be traced in stone mine workers and among the surrounding population. To analyse the impact of crushers, various types of parameters are used, such as the size of dust particles, PM_{10} or $PM_{2.5}$, and amount of formation of oxides of sulphur and nitrogen.

There is an urgent need to study the existence of stone crushers because of the following reasons:

1. To analyse whether the stone crushers are fulfilling the minimum legal requirements laid down for running stone crushers in India.
2. To identify the potential sources of pollution evolving from stone crushers.
3. To prepare environmental safeguards as a part of management plan to mitigate environmental degradations caused by stone mining.
4. To develop a comprehensive environmental monitoring plan to assess the implementation of environmental safeguards.

3 Impact of Stone Mining and Human Health

Human lives are affected directly and indirectly by environmental deterioration. It increases or decreases the physical and non-physical quality of life (Nangi et al. 2020). These effects can also trigger the emergence of various environmental conflicts that can result in human health problems (Fig. 1) and also destroy the fragile ecology of forests, biodiversity, and scenery beauty of the area (Deshar et al. 2013). The problems associated with stone mining and other quarrying may be quite temporary but they can lead to permanent diseases. Stone dust contains chemicals that cause respiratory lung as well as other types of diseases. These diseases are discovered in people working at the site of a rock quarry or in rock transportation because of the continuous inhaling of rock dust. The primary health concern is crystalline silica dust inhalation that settles in the lungs, producing a variety of respiratory and pulmonary disorders, such as TB, bronchitis, silicosis, lung cancer, and other diseases (Chen et al. 2012; Esmaeil et al. 2014; Haque and Singh 2017). According to Chauhan and Sanjeev (2008), in India, 40% of air pollution is the result of dust particles originating from different point sources.

4 Stone Mining in Himachal Pradesh

The state of Himachal Pradesh in India is blessed with fragile ecosystem and biodiversity, and is divided into 12 districts (Fig. 2). It has a diverse range of minerals and rock types (Table 1), which has resulted in a variety of mining activities. Out of different mining activities, stone crushers have

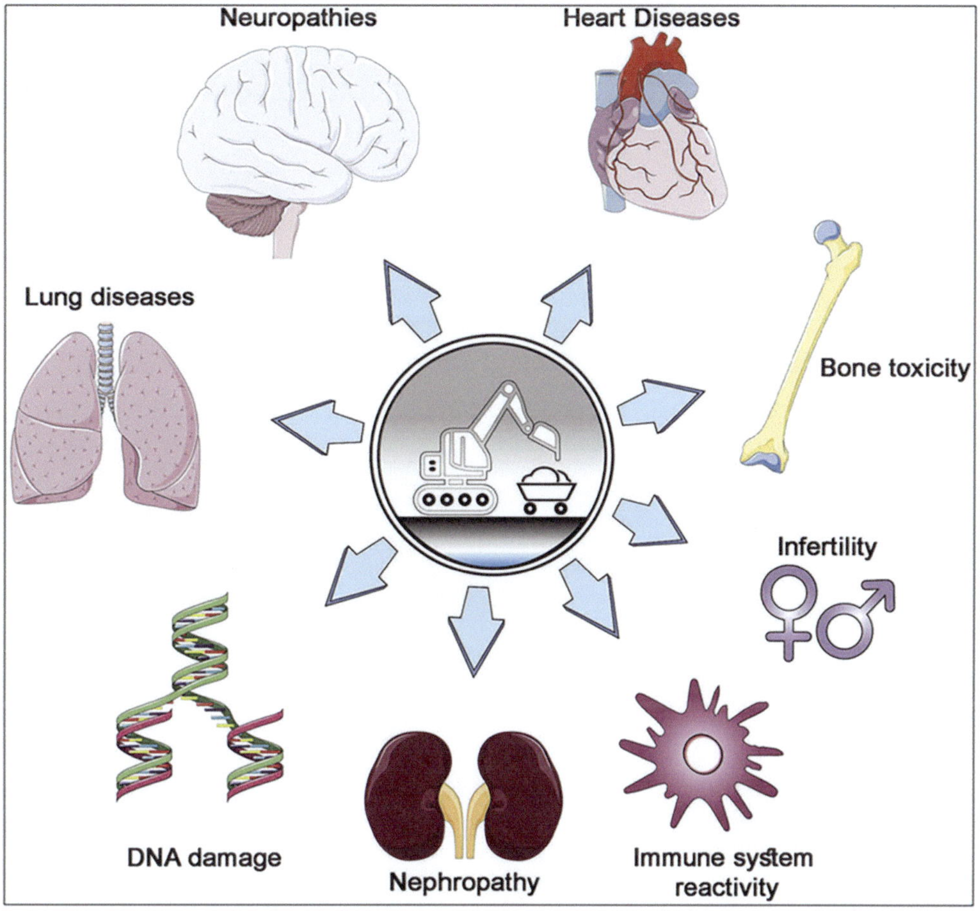

Fig. 1 Impact of mining on human health

occupied a huge area of mining in the state. According to the Himachal Pradesh Mineral Rules 1971 (HPSPCB 2014), mining of building stone, ordinary clay, gravel, ordinary sand, and other types of sandstone are referred to as 'minor minerals' and its excavation is suggested. The excavation of boulder or chalcedony is allowed for mills purposes only, and the mining of lime shell, *kankar* (stone chips), and limestone for kiln for the production of lime and bricks is allowed for building purposes. Mining is also allowed for building materials, such as moram, earth brick, fullers earth metal, slate, and shale. Quartzite and sandstone exploration is allowed for the purpose of construction of metal roads, household utensils, bore well fillings, etc. All these are notified in the Central Government Official Gazette and are declared under (HPSPCB 2014) minor minerals Clause (e) Sect. 3 of the Act (HPSPCB 2014).

4.1 Methodology

In the present work, different data sets of population, diseases, forest cover, and geology were collected from the official sites of the government of Himachal Pradesh, hospitals as well as field surveys. During field visits, the impact of dust on the local environment, local flora, and human health was analysed around the mining sites. For land use and land cover changes, data of advanced remote sensing satellite, Resourcesat-1, was used; the satellite is designed to provide both multi-spectral and panchromatic imagery of the earth's surface. It has three sensors: acquired image was recorded by Linear Imaging and Self-Scanning Sensor (LISS)-III, having a spatial resolution of 23.5 m (Fig. 3). For detecting changes in land use and land cover, images for the years 2005–2015 were analysed. The satellite images were freely downloaded from the Bhuvan platform (https://bhuvan.nrsc.gov.in/). The attributes used in the imageries were further validated by the ground truthing method using the global positioning system (GPS).

5 Stone Mining in Hamirpur

District Hamirpur in Himachal Pradesh lies between latitude 31.67° N and longitude 76.50° E. The total area of the district is 1,118 km^2 and covers about 1.97% of the state. Hamirpur is the district headquarters and is situated at an altitude of 780 height above mean sea level (mamsl).

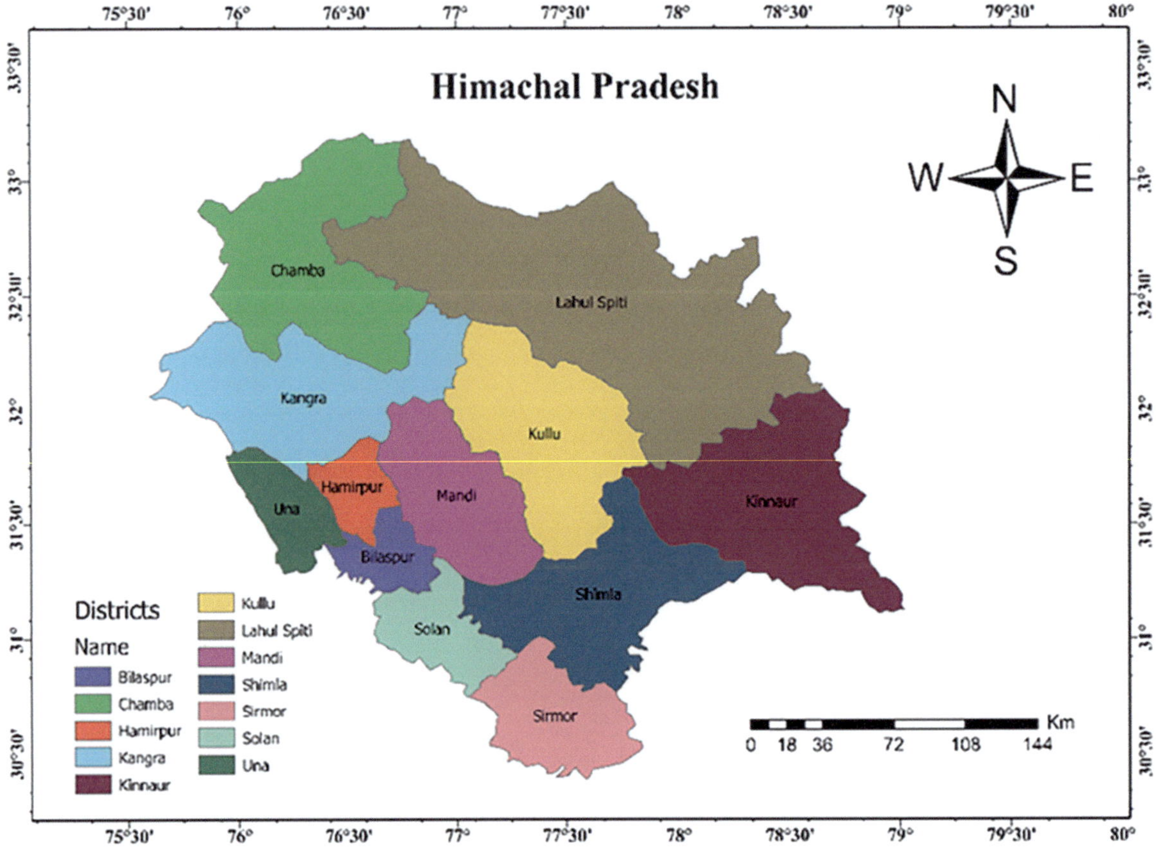

Fig. 2 Location map of Hamirpur in Himachal Pradesh, India

Table 1 Exploitable minerals in different districts of Himachal Pradesh

District	Exploitable minerals
Bilaspur	Limestone, dolomitic limestone, shale, brick earth, minor minerals like sand, stone, and bajri
Kangra	Roofing slate, brick earth, minor minerals, such as sand, stone, and bajri, Chamba limestone, roofing slate, magnesite, and minor minerals, such as sand, stone, and bajri
Kinnaur	Gypsum and minor minerals, such as sand, stone, and bajri Kullu roofing slate, quartzite crystal, mineral water, semi-precious stones and minor minerals, such as sand, stone, and bajri
Lahaul and Spiti	Antimony ore, gypsum, and minor minerals, such as sand, stone, and bajri, Solan limestone, dolomitic limestone, shale, building stone, and minor minerals like sand, stone, and bajri
Una	Silica boulders and minor minerals like sand, stone, and bajri
Mandi	Limestone, rock salt, and minor minerals like sand, stone, and bajri
Hamirpur	Minor minerals like sand, stone, and bajri
Shimla	Limestone, shale, quartzite, and minor minerals like sand, stone, and bajri
Simour	Limestone, barites, gypsum, shale, quartzite, and minor minerals like sand, stone, and bajri

Source Himachal Pradesh Mineral Policy 2013

Stone mining in Hamirpur is not recorded in the literature. But since one of the major rivers of Himachal Pradesh, the river Beas, passes through Hamirpur, the area is rich in sand deposits, hence sand mining is not new for the area. According to government data, there were 18 crushers (Fig. 4), including 15 functional crushers (Table 2), in Hamirpur District in the year 2013 (HPSPCB 2013). According to the latest report of the District Environmental Plan of Hamirpur (DEPH 2021), about 40 crushers were registered in the year 2021 (Tables 2 and 3); however, the field survey had discovered many mining sites apart from the registered ones.

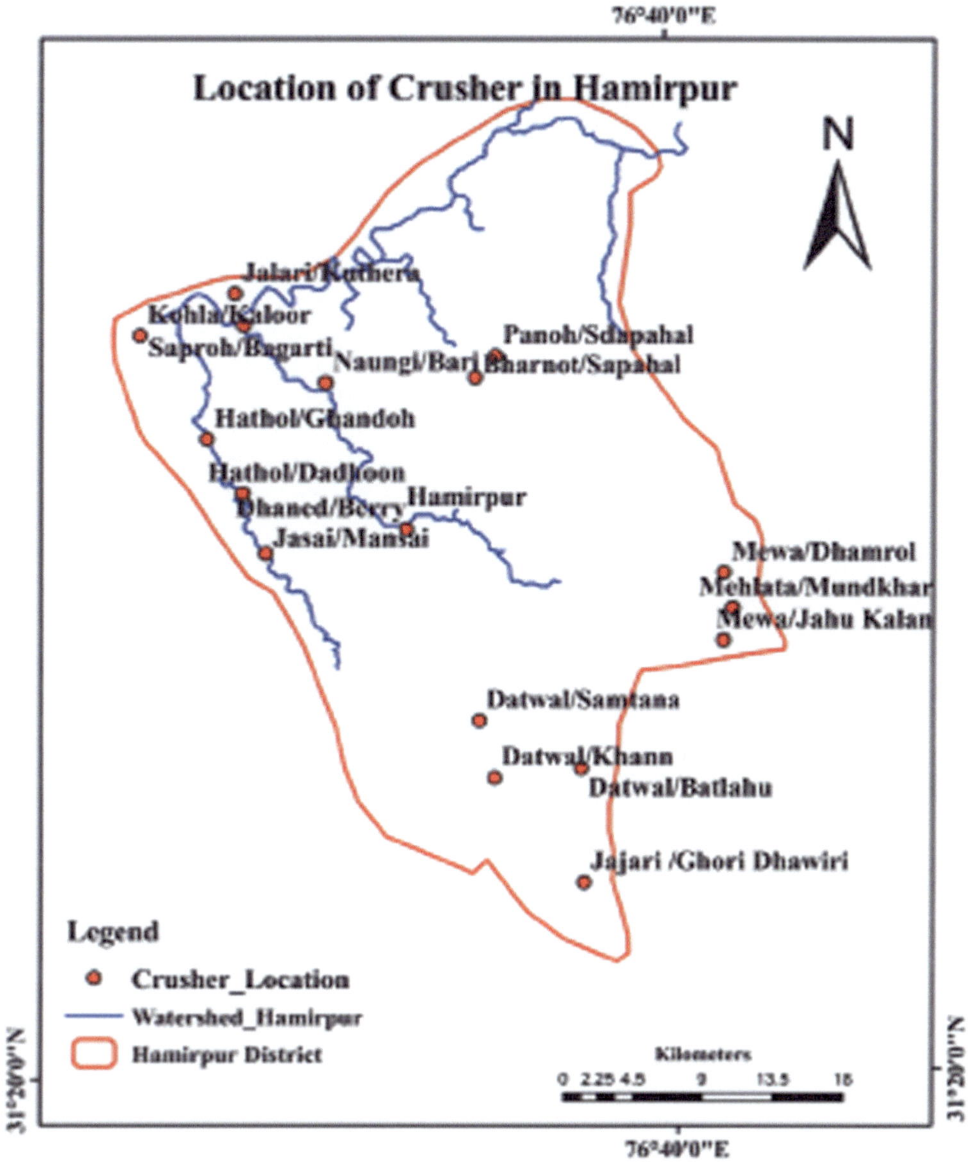

Fig. 3 Location map of stone crushers at district Hamirpur in Himachal Pradesh

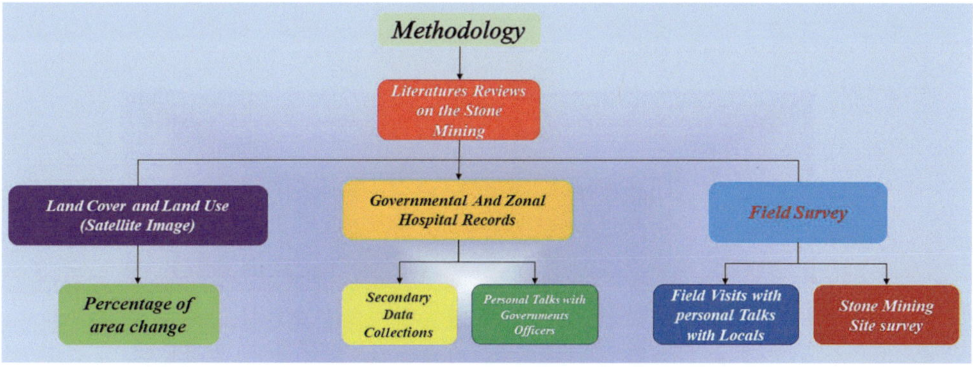

Fig. 4 Flowchart of the methodology used for detecting the impact of stone crushers in the study area

Table 2 Location of stone crushers in Hamirpur

Sr. no.	Name and address	Location (in Mauza, Mohal)	Location coordinates (latitude and longitude)	Area (hectares)	Type (riverbed/ hill slope)	Period in years	Purpose (open sale/stone crusher)	Status (operational/ non-operational) if non-working, reason thereof with the date of non-working
1	Sh. Raj Kumar, Prop.: M/s Jai Bhole Shankar Stone Crusher, VPO Samtana, Teh. Barsar, Dist. Hamirpur	Datwal/Samtana	31° 32′ 25″ N/ 76° 34′ 44″ E	4-96-62	Riverbed	5 years 16.01.2012–15.01.2017	Stone crusher	Non-functional
2	Smt. Saroti Devi, Prop.: M/s Jai Baba Stone Crusher, VPO Dhangota, Teh. Barsar, Dist. Hamirpur	Datwal/Batlahu	31° 30′ 42″ N/ 76° 36′ 15″ E	4-44-00	Riverbed	5 years 23.5.2015–22.5.2020	Stone crusher	Functional
3	Sh. Subhash Chand, S/o Sh. Mukadam Singh VPO Laharli, Teh. Barsar, Dist. Hamirpur	Jajari/Ghori Dhawiri	31° 26′ 42″ N/ 76° 36′ 18″ E	19-34-61	Riverbed	15 years 2.11.2011–01.11.26	Stone crusher	Functional
4	Sh. Ajay Kumar, S/o Shri Kuldeep Singh Rana, VPO Dhangota, Teh. Barsar, Dist. Hamirpur	Datwal/Khann	31° 30′ 24″ N/ 76° 32′ 42″ E	3-83-26	Riverbed	5 years 10.02.2012–09.02.2017	Open sale	Non-functional/ suspended since 5.8.2013 for want of environment clearance
5	Sh. Bharat Bhushan, S/o Sh. Ravinder Nath, VPO Dhaneta, Teh. Nadaun, Dist. Hamirpur	Jasai/Mansai	31° 38′ 22″ N/ 76° 23′ 37″ E	1-65-71	Riverbed	5 years 05.08.2015–04.08.2020	Open sale	Functional
6	Sh. Ashok Thakur, Prop. Maheshwar Stones Crushing and Washing Unit, VPO Kuthera, PO Jalari, Teh. Nadaun, Dist. Hamirpur	Jalari/Kuthera	31° 47′ 30″ N/ 76° 22′ 30″ E	4-03-97	Riverbed	15 years 13.3.2014–12.3.2029	Stone crusher	Functional
7	M/s Shivalik Stones Crushers, Prop. Sh. Dharambir Thakur, VPO Rangas, Dis	Naungi/Bari	31° 44′ 20″ N/ 76° 26′ 05″ E	4-05-33	Riverbed	5 years 17.1.2012–16.1.2017	Stone crusher	Non-functional
8	Shri Vijender Jain, S/o Shri Sudershn Jain, Village Seri, PO & Teh. Nadaun, Dist. Hamirpur	Saproh/Bagarti	31° 46′ 23″ N/ 76° 22′ 50.5″ E	3-47-58	Riverbed	5 years 20.10.2015–19.10.2020	Stone crusher	Functional
9	Sh. Lekh Raj Sharma, S/o Sh. Jaishi Ram, Village Badehra. PO Batran, Teh. Nadaun, Dist. Hamirpur	Hathol/Dadhoon	31° 40′ 28.7″ N/76° 22′ 43.4″ E	6-29-34	Riverbed	5 years 28.12.2015–27.12.2020	Stone crusher	Functional

(continued)

Table 2 (continued)

Sr. no.	Name and address	Location (in Mauza, Mohal)	Location coordinates (latitude and longitude)	Area (hectares)	Type (riverbed/ hill slope)	Period in years	Purpose (open sale/stone crusher)	Status (operational/ non-operational) if non-working, reason thereof with the date of non-working
10	Sh. Satish Kumar, S/o Sh. Bidhi Chand, Village Baloh, PO Jallan, Teh. Nadaun, Dist. Hamirpur	Hathol/Ghandoh	31° 42′ 24.4″ N/76° 21′ 19.34″ E	4-79-80	Riverbed	5 years 5.5.2016–4.5.2021	For free sale	Functional
11	Sh. Gian Chand, S/o Sh. Chunni Lal, VPO Bharmoti, Teh. Nadaun, Dist. Hamirpur	Kohla/Kaloor	31° 46′ 03.8″ N/76° 18′ 40.8″ E	14-58-14	Riverbed	15 years 16.6.2016–15.6.2031	For stone crusher	Functional
12	Smt. Ashlata Devi, W/o Sh. Nand Lal, Vill. Manjhot, PO Ropa, Teh. and Dist. Hamirpur	Panoh/Sdapahal	31° 45′ 12.3″ N/76° 32′ 57.2″ E	1-06-73	Hill slope	3 years 2.3.2016–1.3.2019	Stone crusher	Functional
13	M/s Sanjay Chouhan Stones Crusher-2, Jahu, Teh. Bhoranj, Dist. Hamirpur	Mewa/Jahu Kalan	31° 35′ 08″ N/ 76° 42′ 07″ E	17-30-76	Riverbed	15 years 20.10.2011–19.10.2026	Stone crusher	Functional
14	M/s Sanjay Chouhan Stones Crusher, Jahu, Teh. Bhoranj, Dist. Hamirpur	Mehlata/Manoh Bulla and Behrwin Jattan	31° 69′ 40″ N/ 76° 40′ 96″ E	7	Riverbed	15 years 13.06.2013–12.06.2028	Stone crusher	Functional
15	M/s Sanjay Chouhan Stones Crusher, Jahu, Teh. Bhoranj, Dist. Hamirpur	Mewa/Dhamrol	31° 37′ 30″ N/ 76° 42′ 10″ E	1-50-46	Riverbed	15 years 13.06.2013–12.06.2028	Stone crusher	Functional
16	Sh. Purshotam Chand, Prop M/s Himachal Stone Crusher, Mundkhar, Teh. Bhoranj, Dist. Hamirpur	Mehlata/ Mundkhar	31° 36′ 15″ N/ 76° 42′ 30″ E	11-39-56	Riverbed	15 years 19.08.2015–18.08.2030	Stone crusher	Functional
17	M/s Jagdamba Crusher Co VPO Massiana, Teh. and. Dist. Hamirpur (HP)	Dhaned/Berry	31° 39′ 10″ N/ 76° 29′ 15″ E	1-75-85	Riverbed	5 years 18.3.2016–17.3.2021	Stone crusher	Functional
18	Sh. Parveen Kumar, Prop M/s Mahabir Stones Crusher, VPO Bhatera, Teh. and Dist. Hamirpur	Bharnot/Sapahal	31° 44′ 27″ N/ 76° 32′ 06″ E	4-19-23	Riverbed	15 years 12.7.2011–11.7.2026	Stone crusher	Functional

Source Himachal Pradesh Geological Wing 2013

Table 3 Functional mining details by DEPH 2021

Details of mining data 2021	Existing mining functions
Type of mining activity	Sand, stone, and bajri
No. of licensed mining operations in the district	40
Area covered under mining in the district	246-12-21 hectare

Source District Environment Plan Hamirpur 2021

Table 4 Number of patients registered in Nadaun Civil Hospital of Hamirpur District from 2011 to 2017

Year	No. of patients
2011	1,985
2012	2,132
2013	3,580
2014	3,880
2015	4,480
2016	5,598
2017	5,998

Source Civil Hospital, Nadaun Block

5.1 Impact of Stone Mining and Human Health in Hamirpur

Field visits and local health survey discovered detrimental effects on the health of crusher workers and local communities, such as respiratory diseases, also discovered in the workers of other states of India. Other diseases discovered include headache, diarrhoea, irritation in the eyes and skin, fever, chest pain, and lung ailments (Ambastha and Haritash 2022; Esmaeil et al. 2014; Manisalidis et al. 2020; Singh et al. 2009, 2017).

There are several point sources of dust and noise pollution, both continuous and intermittent. Vibratory screens, pulleys, and convoy belts of stone crusher units are the most predominant sources of continuous noise pollution (CPCB 2009; HPSPCB 2014).

In our case study, health data was collected from 2011 to 2017 from the government hospital Nadaun, Hamirpur. The hospital data revealed that number of patients registered with crushers-related diseases had increased every year, although the records were not disease-specific (Table 4).

5.2 Land Use and Land Cover Changes

In order to study the impact of stone mining in the area, the land use and land cover map was prepared for the period 2005–2015. The LISS-III satellite images were analysed using the ArcMap 10.4 application. Supervised classification was carried out for the collected data for the period 2005–2015, in which different classes were assigned that included water bodies, barren land, agricultural land, forest land, and urban and rural settlements (Figs. 5 and 6).

Table 5 shows changes in land use and land cover between 2005 and 2015. The analysis demonstrated that barren land increased by 5.86%, and agricultural land and forest land decreased by 4.28% and 3.80%, respectively. An increase of 0.95% was observed in the urban and rural settlement pattern, while the area of water bodies decreased by 0.08%. No signatures of stone mining were discovered in the mining areas of Hamirpur District in the year 2005, but in the year 2015, stone mines occupied around 0.01% of the area. All mining plants are situated on the riverbeds of river Beas in Himachal Pradesh.

Field surveys discovered that all stone crusher plants were located along the river bank or the stream. The riverbed and hill slope of the area were exploited excessively due to the extraction of materials for stone crusher units. The operational stone crusher units released dust that affected local vegetation and agriculture fields (Figs. 7 and 8).

5.3 Unscientific and Haphazard Mining

Most of the stone mining plants in the Hamirpur area are operating unscientifically, because riverbed mining is causing severe environmental problems, such as the degradation of aquatic habitat ecosystems, which requires instant attention and remediations. In recent years, the impact of riverbed mining (sand, gravel, and boulder) on the environment has become more prominent in the area and has come under increased scrutiny (HPSPCB 2014; Kamboj et al. 2017).

5.4 Mitigation Measures

Many studies have demonstrated and summarised the major effects of excessive sand, stone chips, and stone mining and their impact on the environment, flora, and fauna. Thus, there is an urgent need to mitigate the problems related to mining to conserve different components of the environment.

(a) Suggestive measures
   1. Labourers should be given sufficient safety equipment to manage the problem of dust, such as masks, protective head gears, hand gloves, and protective eye glasses.
   2. The government and stone crusher companies should work together to educate workers and encourage them to get regular medical consultations. This helps them to lower the risk of respiratory diseases.
   3. All workers in the mining industry should be enrolled in the government's health schemes and take health

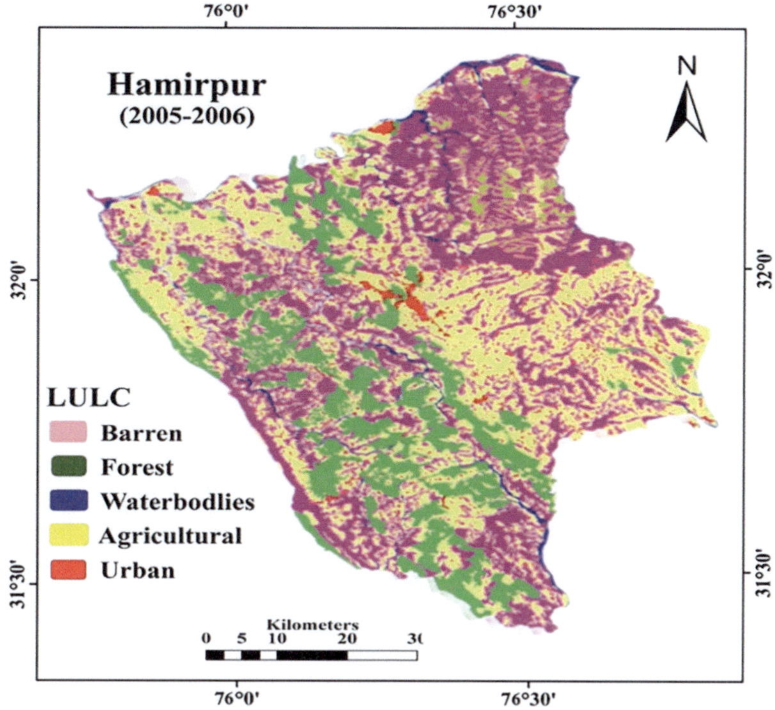

Fig. 5 Land use and land cover map of the study area during 2005–2006

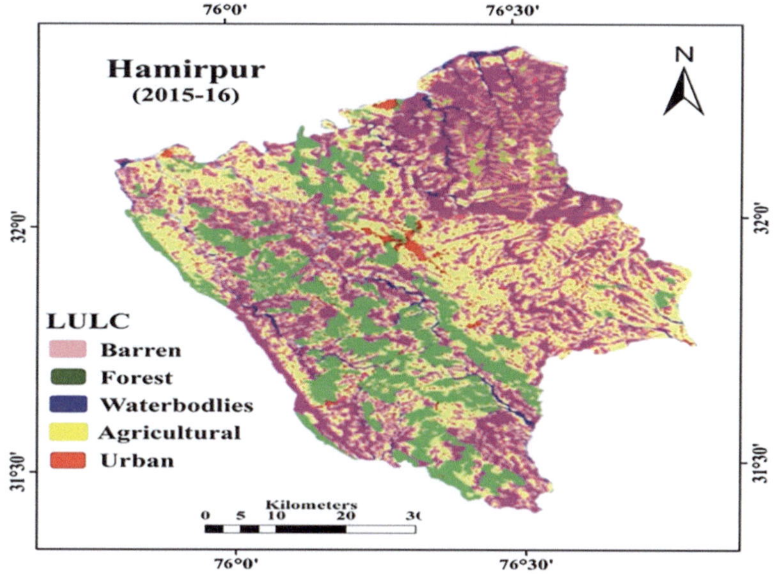

Fig. 6 Land use and land cover map of the study area during 2015–2016

Table 5 Observed changes in land use and land cover of the study area between 2005 and 2015

Name	Year 2005 (area in sq. km)	Percentage (%)	Year 2015 (area in sq. km)	Percentage (%)	Percentage difference
Urban/rural	26.39	2.36	37.02	3.31	0.95
Agriculture	447.69	40.04	399.89	35.77	−4.28
Forest	265.72	23.77	223.28	19.97	−3.80
Barren	332.68	29.76	398.16	35.61	5.86
Water bodies	45.52	4.07	44.68	4.00	−0.08
Mining	–	–	0.08	0.007	0.01

Source bhuvan.nrsc

Fig. 7 Operational units in stone mining site

Fig. 8 Vegetation around the stone mining site

insurance. The government should analyse the speed or direction of wind/air of stone crusher units to track the movement and direction of stone dust.

4. River water quality should be monitored on a regular basis to avoid deterioration of aquatic habitat.
5. Environmental Impact Assessment (EIA) should be made mandatory for every proposed crusher unit; this helps in decision-making and analysis of environmental consequences.

(b) Preventive measures

1. Stone mining should be restricted to operate during monsoonal months.
2. Dimensions of stone mining area, depth of excavation pit, and size of boulders should be decided according to their vulnerability.
3. Empty pits should be filled with solid material of mining waste during the rainy season.
4. Stone mining operations should be avoided during high wind speeds and harvest seasons.

6 Conclusions

- It is observed that large stone crusher units release large-scale dust around the mining area. These large plants must be equipped with water sprayers to settle down the dust and prevent it from resuspension.
- Transportation and unloading/loading of raw material from open stocking of crushed stone products of different sizes release huge amount of dust. The paths should be moistened with water sprays.
- It was observed during local field visits that dust directly affects local flora and fauna, agriculture, and human health.

- The number of operational stone crushers in District Hamirpur exceeds the numbers registered in government records, showing that most of the stone crushers are operating illegally.
- Disease-specific records in hospitals should be maintained to identify the frequency of respiratory diseases resulting from stone mining.
- Most of the stone-crushing units in the area are located on the bank of river Beas. During heavy downpour, all their dust and debris are carried to the river that results in serious environmental and ecological problems.
- Enormous clouds of dust are observed around crushing unit areas that remain suspended in the air and on vegetation even after the operation is over.
- Riverbeds are heavily exploited by crushing units, leading to riverbed erosion.
- Stone mining management and local authorities of the areas do not create awareness among the local population regarding their medical consultations.
- In a local civil hospital, only the entry of the patients is recorded. We did not find any records related to the type of diseases suffered by the patients.
- At the time a mining lease is given to stone mining agencies, a comprehensive policy framework must ensure that there should be minimum environmental degradation. It was discovered that policy guidelines were not followed by more than half of the registered stone mining units of the area.

References

Ambastha, S.K. and Haritash, A.K. (2022). Emission of respirable dust from stone quarrying, potential health effects, and its management. *Environmental Science and Pollution Research*, **29(5)**: 6670–6677. https://doi.org/10.1007/s11356-021-16079-4

Ambastha, S.K., Haritash, A.K., Sivacoumar, R., Jayabalou, R., Subrahmanyam, Y.V, Jothikumar, N., Swarnalatha, S., Ahmad, A.F., Sharma, H.K.J., Ahmad, R.M. and Rao, R.J.J. (2014). Impact of mining activities on various environmental attributes with specific reference to health impacts in Shatabdipuram, Gwalior, India. *Environmental Science and Pollution Research*, **43(6)**: 81–87.

Central Pollution Control Board (CPCB). (2009). Comprehensive Industry Document Stone Crushers. February. CPCB, New Delhi, India.

Chaurasia, S., Singh, R. and Pathak, V. (2009). Environmental study of stone crusher. *Cough*, **35(20)**: 15–57.

Chen, W., Liu, Y., Huang, X. and Rong, Y. (2012). Respiratory diseases among dust-exposed workers. *In:* Respiratory Diseases (Eds. M. Ghanei), pp.132–150 xxx IntechOpen, London. https://doi.org/10.5772/32357

Chauhan, A. and Sanjeev (2008). Impact of dust pollution on photosynthetic pigments of some selected trees grown at nearby of stone-crushers. *Environment Conservation Journal*, **9(3)**: 11–13.

Deshar, B.D., Manisalidis, I., Stavropoulou, E., Stavropoulos, A., Bezirtzoglou, E., Upadhyaya, M., Haque, M. and Singh, R. (2013). A study on mining industry pollution in Chapagaon, Nepal. *Applied Ecology and Environmental Sciences*, **2(4)**: 806–810. https://doi.org/10.3389/fpubh.2020.00014

Dhar, B.B. and Ahmad, M. (1993). Impact of mining and processing activities on surrounding environment – A case study. *Journal of Mining Research*, **2(2)**: 34–41.

District Environment Plan of Hamirpur (DEPH). (2021). DEPH, HP. District environmental plan-Hamirpur, Govt. of Himachal Pradesh. https://hphamirpur.nic.in/document/district-environment-plan-hamirpur/

Esmaeil, N., Gharagozloo, M., Rezaei, A. and Grunig, G. (2014). Dust events, pulmonary diseases and immune system. *American Journal of Clinical and Experimental Immunology*, **3(1)**: 20–29.

Haque, M. and Singh, R. (2017). Air pollution and human health in Kolkata, India: A case study. *Climate*, **5(4)**: 77. https://doi.org/10.3390/cli5040077

Himachal Pradesh State Pollution Control Board (HPSPCB). (2014). Comprehensive industry document stone crushers. *Safety and Health at Work*, **19(4)**: 1–20. https://doi.org/10.4172/21626359.1000

Himachal Pradesh State Pollution Control Board (HPSPCB). (2019). Annual report, 96–107. https://hppcb.nic.in/Publications/AR-2019-20.pdf

Kamboj, V., Kamboj, N. and Sharma, S. (2017). Environmental impact of river bed mining – A review. *International Journal of Economic and Environment Geology*, **7(1)**: 504–520.

Kamboj, V., Kamboj, N. and Sharma, S. (2018). Environmental impact of river Bed mining – A review. *International Journal of Scientific Research and Reviews*, **7(1)**: 504–520.

Karmakar, H.N. and Das, P.K. (2012). Impact of mining on ground and surface waters. *In:* 4th international mineral water association congress (pp., 187–198f).

Lerman, S.L. and Darley, E.F. (1975). Particulates. *In:* Responses of Plants to Air Pollution, A Volume in Pergamon Studies in Neuroscience (Eds. Brian Mudd, J. and Kazlowski, T.T.), pp. 141–158. Academic Press, New York, NY.

Manisalidis, I., Stavropoulou, E., Stavropoulos, A. and Bezirtzoglou, E. (2020). Environmental and health impacts of air pollution: A review. *Frontiers in Public Health*, 8, 14. https://doi.org/10.3389/fpubh.2020.00014

Manzoor, J. and Khan, M. (2020). Environmental perspective of stone crushing and quarrying : A case study from Poonch district of Jammu and Kashmir, India. *Current World Environment,* **15 (1)**: 68.

Mehta, P.S. (2002). The Indian mining sectors: Effects in the environment & FDI inflows. *In:* Conference paper, *Conference on foreign direct investment and the environment* (pp., 7–8).OECD Headquaters, **17**. https://www.oecd.org/env/1830307.pdf

Nangi, M.G., Saparina, T., Lestari, S.A. and Juslan. (2020). The impact of stone mining on community health around mining in North Moramo sdistrict. *In:* Proceedings of the 4th International Symposium on Health Research (ISHR 2019), Atlantis press, pp. 87–89. https://doi.org/10.2991/ahsr.k.200215.017

Pal, S. and Mandal, I. (2019a). Impacts of stone mining and crushing on environmental health in Dwarka river basin. *Geocarto International,* **36(4)**: 392–420. https://doi.org/10.1080/10106049.2019.1597390

Pal, S. and Mandal, I. (2019b). Impact of aggregate quarrying and crushing on socio-ecological components of Chottanagpur plateau fringe area of India. *Environmental Earth Sciences*, **78**: 661. https://doi.org/10.1007/s12665-019-8678-1

Patil, M.A. (2001). Environmental management scenario in stone crusher industry sector and cleaner production possibilities. *TERI Information Monitor on Environment Science*, **6**: 83–92.

Singh, K.K., Negi, H.S., Kumar, A., Kulkarni, A.V., Dewali, S.K., Datt, P., Ganju, A. and Kumar, S. (2017). Estimation of snow accumulation on Samudra Tapu glacier, Western Himalaya using airborne ground penetrating radar. *Current Science*, **112(6):** 1208–1218. https://doi.org/10.18520/cs/v112/i06/1208-1218

Singh, G., Pal, A. and Khoiyanbam, R.S. (2009). Impact of mining on human health in and around Jhansi, Bundelkhand region of Uttar Pradesh, India. *Journal of Ecophysiology and Occupational Health*, **9(1-2):** 47–54.

Talwar, P., Kumar, J. and Loshali, D.C. (2017). Environmental effect of stone crusher industry on riverbed using geo-spatial techniques : A case study in Pathankot district, Punjab. *International Journal for Research in Applied Science & Engineering Technology* (*IJRASET*), **5(viii):** 2026–2032.

Water Quality Index Review—Trends and Views

Yarly E. Mosquera-Torres

1 Introduction

Without water, life would not exist, because plants and crops feed on the minerals found in the soil, but in order to assimilate them they must be dissolved. Plants absorb water with dissolved minerals through their roots, and through the process of photosynthesis transform them into food substances for other living beings, including humans (Auge 2007). On the one hand, plants, since ancient times, have established their populations in the areas near the rivers and provide food security, and on the other, they have become the direct and indirect collectors of human activity. These situations resulted in population growth, and the lack of adequate wastewater evacuation systems aggravated contaminations present in water sources, affecting both quality and quantity of water. Both anthropogenic activity and the natural dynamics of water bodies are crucial in explaining and protecting water quality (Tibebe et al. 2022). Therefore, freshwater sources, including the environmental services they provide, are threatened, reducing the self-purification and assimilation capacity of rivers, which lead to a decrease in water (Kumar et al. 2022). Hence, it is important to permanently monitor these sources as a vital task in the preservation and comprehensive management of water resources, and thus prevent this contamination from becoming a serious local, national, and/or global problem.

This subject is being investigated since 1965, and different authors have studied the state of water sources and designed methodologies that would allow establishing an index to determine their quality. Till the year 2002, the studies in relation to the water quality index (WQI) remained constant, and since the year 2009, studies on this subject have increased, reaching its peak in 2019. Maximum research in this field has taken place in China, the United States, Russia, Brazil, Japan, the United Kingdom, India (Khatri et al. 2020), Spain (Pascual Aguilar et al. 2019), Egypt (Abdel-Satar et al. 2017) and Canada (Kachroud et al. 2019) and have used the indices to determine the quality of water in their water sources.

During the above-mentioned period, WQIs were established for water bodies in different countries with the aim of determining their quality and establishing their current status. In the order of ideas, among the indices that we can list are Prati, Dalmatia, WQI-National Sanitation Foundation (NSF), Montoya, and CCME-WQI, developed in the rivers of Miami, Oregon (Castro et al. 2014), Idaho (Samboni et al. 2007), and Des Moines (Fernández et al. 2004a, b) in the United States. WQIs were established in the Nile River (Abdel-Satar et al. 2017), in the Mediterranean Rivers (Bustamante Toro et al. 2008), as well as the Amajac River (Álvarez et al. 2008), the rivers of the Lerma-Chapala Basin, and the Tecoluth River (Arriaga-Gaona et al. 2009) in Mexico; the Naranjo River in Cuba (Tunas 2014); and the Chumbao River in Peru. These were used in the evaluation of 43 water quality parameters. In addition, the quality indices were also proposed by WQI-Dinuis, ISQA-Spain, IAP-Brazil, ICAUCA-Colombia and MINAM-Peru (Choque_Quispe, et al. 2022) as well as in the Pita and Paute Rivers in Ecuador (Simbaña-Farinango et al. 2019; Bajaña et al. 2022).

Other developed indices are the Yungas Biotic Index (IBY-4), based on four parameters. It explains the appearance of megaloptera, plecoptera, trichoptera, and elmidae at the target site and could achieve the best overall performance in the study region regarding Andean tropical currents (Dos Santos et al. 2011). The River Recreation Index (RRIM) was developed on the Nakdong River in South Korea (Boyacioglu 2007). The proposed Tr-WQI index for the Tigris River (Iraq) considered 20 water quality parameters (Abed et al. 2022).

Y. E. Mosquera-Torres (✉)
Bloque 3 Laboratorio de Hidráulica, Universidad Tecnológica del Chocó, Grupo de Investigación Ingeniería Civil—IC, Barrio Nicolás Medrano, Quibdó, Colombia
e-mail: yamostorres@hotmail.com

J. Marrugo-Negrete et al. (eds.), *Remediation and Carbon Capture in Mining Sector*, https://doi.org/10.1007/978-3-031-87883-1_5

In Colombia, indices on the Cauca River, ICAUCA index (CVC_UNIVALLE 2004), stand out along with indices on the Quindío River (Bustamante Toro et al. 2008), the Córdoba River (Fontalvo et al. 2018), the Ayurá creek (González et al. 2012), the Ranchería River (Pérez et al. 2018), the Cali and Meléndez Rivers (Reinoso-Florez et al. 2008) etc. as well as several water contamination indices (WCI) (Chavarro and Gelvez Bernal 2016). In the Department of Chocó, studies were carried out on the quality of water in rivers, such as the San Juan and Cabí Rivers (Ríos-Tobón et al. 2017; Yimmy Montoya et al. 2011).

2 Water Quality Indices

2.1 Definition of a Water Quality Index

A WQI basically consists of a simple expression or a more or less complex combination of a number of parameters with which water quality is characterised (Lino et al. 2009; Gómez Delgado 1995). It determines the general physicochemical conditions of the quality of a body of water and, to some extent, allows recognising contamination problems at a given point for a specific time interval. It allows representation of the general state of water and the possibilities or limitations for certain uses based on selected variables through weighting and aggregation of physical, chemical, and biological variables (Lino et al. 2009).

The WQI and contamination index (ICO) reduce a large number of parameters to a simple expression of easy-to-simplify interpretation of monitoring data among technicians, environmental managers, and the general public (Lino et al. 2009). Additionally, the WQI is a tool that allows identifying the water quality of a surface or underground body at a given time (Caho-Rodríguez and López-Barrera 2017; Chavarro and Gelvez Bernal 2016; Ríos-Tobón et al. 2017; Yimmy Montoya et al. 2011). A general analysis of water quality at different levels can be carried out through WQI and vulnerability of the body against potential threats can be determined (Agrama 2012). The indices constitute a useful tool in the diagnosis, surveillance, control, and integrated management of water resources, since through their definition, actions could be taken to control impact on the water source.

Thanks to the disaggregation of indices, specific contamination situations could be evaluated using some of them, monitoring only the variables that are of concern. This fact allows, on the one hand, cost savings, and on the other, concrete results in relation to the problem under study (Fernández et al. 2004a, b; Posada et el. 2008). In addition to the above, the WQI is a useful tool that allows identifying the sources of contamination and the space–time variation that they may have, which is very important in the framework of water quality management (Lopes et al. 2021). This allows the implementation of corrective measures to prevent and control the deterioration of water bodies (Valcarcel et al. 2009), in addition to providing a scientific justification for the planning and management of water use (Giao et al. 2021).

2.2 Background

Table 1 summarises in chronological order some of the studies carried out in relation to water contamination rates.

2.3 Classification of Water Quality Indices

According to Samboni et al. (2007), the WQI can be organised into 10 categories within four groups as shown in Table 2.

2.4 Uses of Water Quality Indices

Each WQI aims to reduce different parameters to be taken into account to a single variable when evaluating a water source, figuring, of course, the regulations applicable to each region or country as well as advances in technology. Because water quality varies both spatially and temporally in relation to its uses, this aspect must also be considered and the variables that are considered in measuring WQI must be chosen. Thus, the number of variables to analyse or measure should not be so few that they are not decisive when establishing the quality of the source, nor should they be too many that it becomes impossible to evaluate the index.

According to Ott (1978) and Torres Vega (2009), the use of WQIs can be grouped as follows:

- *Resource management:* Provides information to people who make decisions about water resources.
- *Classification of areas:* The state of a water source can be compared in different areas.
- *Application of regulations:* It can be determined if the existing environmental regulations and policies are being passed.
- *Trend analysis:* Over a period, one can show whether the quality of the source is increasing or decreasing.
- *Public information*: To raise awareness and environmental education.
- *Scientific research*: Simplifies a large amount of data and in this way can provide a view of the environmental phenomena present in the water source.

In this sense, the sources can be classified according to WQIs as indicated in Table 3.

Table 1 Background of contamination indices

Year	Author	Description
At the international level		
1965	Horton	Generation of a unified methodology for calculating WQIs: selection of characteristics, establishment of a scale, and weighting of various characteristics
1970	Brown et al.	He developed the general index of water quality
At the international level		
1971	Deininger and Landwehr	Index similar to that of Brown's and contains 12 parameters for surface water and 14 for groundwater. The variables are OD, FC, pH, BOD, concentrations of nitrate, phosphate, phenol, dissolved solid, temperature, turbidity, colour, and hardness
1971	Prati	The authors proposed an index based on water quality standards by transforming contaminant concentrations into contamination levels. The variables used were pH, BOD, chemical oxygen, chemical oxygen demand (COD), DO, concentrations of permanganate, ammonium, nitrate, chloride, iron, manganese, alkyl benzene, sulfonate, carbon chloroform layers, and suspended solids
1972	Dinius	He proposed an ICA made up of nine microbiological variables
1978	Ott	He presented a detailed discussion of environmental index theory and its development
1980	Deininger	Improved the general index of water quality
1983	Bhargava	The author introduced variables focussed more on the pollution load
1985	Tiwary and Mishra	The authors used the principles of Horton and Brown, modified them using normative values
1987	Dinius	He developed another index using the subscripting methods introduced by Dalkey in 1967
1989	House and Newsome	They consider that the use of indexes could quantify water as 'good' and 'bad'
1990	Canadian Council of Environment Ministers	It presented an index developed by the Water Quality Guidelines Working Group
1991	Giordano et al.	The community developed the Universal Water Quality Index (UWQI), which has 12 variables: Total coliforms, pH, DO, BOD_5, total phosphorus, cadmium, cyanide, mercury, selenium, arsenic, fluoride, and nitrates
1995	The strategic assessment of Florida's environment (SAFE)*	It formulated a special index for Florida
1996	Paul	Canada's British Columbia Index (BCWQI) was developed. This is a nonlinear index
1996	Miami Lower Basin Improvement Program	He developed the WEP WQI index that encompasses physical, chemical, and biological aspects. The scale was expressed as excellent, good, fair, and poor
2004	Said et al.	The authors developed a new index by applying logarithmic aggregation
2005	Debels et al.	The water body monitoring programme considered the WQSA. An extended-WQI and a simplified-WQI were developed
2007	Icaga	The calculation is based on fuzzy logic, which allows the use of subjective and non-quantitative data, such as odour
2007	Samboni et al.	The first index to be applied in Mexico was the INDIC-SEDUE, based on the one developed by Diniuis
2010	Bedregal et al.	This has been obtained using the results of the physical–chemical parameters: pH, oxygen (dissolved), conductivity, biochemical oxygen demand, total suspended solids, total phosphorous, nitrate, and ammonia
2011	Bharti and Katyal	Evaluates water quality around mining sites in the United States, identifying changes in water quality over time and space
2011	Karbassi et al.	The factors like dissolved oxygen (DO), fecal coliform (*F-coli*), and biological oxygen demand (BOD) when compared with NSF-WQI
At the local level		
1997	Ramírez et al.	The ICOs were based on the multivariate results in the monitors of the Colombian oil industry. Within these indices, the following stand out: ICOMI, ICOMO, ICOSUS and ICOTRO
1999	Ramírez et al.	The oil industry added the ICOTEMP and ICO pH indices

(continued)

Table 1 (continued)

Year	Author	Description
2004	CVC Univalle	The regional environmental authority (CVC) develops the ICAUCA index for the Cauca River
2004	Fernández et al.	Development of a computer tool for the analysis and assessment of water quality ICATEST

BOD5: biological oxygen demand in 5 days.
Source Prati (1971); Brown, et al., (1970); Dinius, (1972); CVC-UNIVALLE (2004); Sousa et al. (2008); Debels et al. (2005); Bharti and Katyal (2011); Kachroud et al. (2019); Castro et al. (2014); Samboni et al. (2007); Fernandez and Solano (2007); Ramírez et al. (1997); Chávez L. (2010); Jiménez and Vélez (2006); Madsen et al. (2018); Fernández et al. (2004a, b); Dutta et al. (2022).
[a]Florida Environmental Assessment Strategy.

Table 2 Classification of water quality indices

Group	Indicator	Description
I: Applies to tensioners	Indicators at the source Indicators at a point other than the source	They report water quality generated by tensors in discrete sources They report quality of the water generated by diffused sources
II: Measure of stress capacity	Simple measurement indicators Indicators based on criteria or standards	They include many attributes and individual components of water which can be used as indicators of its quality They correlate the measures of water quality with the standard or normal levels that have been determined for the preservation and adequate use of water
III: Measurement in lakes	Multi-parameter indices Empirical multiparameter indices Indicators for lakes	They are determined by the collective or individual opinion of experts They are established by the use of statistical properties of water quality measurements Specifically developed for this type of systems
IV: Takes into account the consequences	Aquatic life indicators Water use indicators Perception-Based indicators	Based on different tolerance reactions of aquatic biota to various contaminants and conditions They carry out the evaluation of water with respect to its uses They are determined by public opinion and the uses of water bodies

Adapted from Castro et al. (2014); Giacomett and Bersosa (2006); Zapata et al. (2006).

Table 3 Classification of sources according to water quality indices

Code	WQI							
	WQI-NSF	WQI-Dinuis	DWQI	UWQI	ISCA	IAP	ICA rojas	ICAUCA
1	Very bad quality	Unacceptable consumption	Poor	Poor	Cannot be used	Lousy	Very bad	Lousy
2	Poor quality	Doubtful for consumption	Marginal	Marginal	Recreation and cooling	Bad	Inadequate	Inadequate
3	Medium quality	Necessary purification treatment	Regular	Regular	Human consumption with special treatment, irrigation, industrial	Regular	Acceptable	Acceptable
4	Good quality	Minor treatment required	Good	Good	Human consumption with conventional treatments	Good	Good	Good
5	Excellent quality	Minor treatment required	Excellent	Excellent	All uses	Optimal	Optimal	Optimal
6	–	Does not require treatment for consumption	–	–	–	–	–	–

Source Kachroud et al. (2019); Castro et al. (2014); Torres Vega (2009); Espejo (2010)

Table 4 Aggregation averages

Method	*Equation*	*Meaning of terms*
Weighted average	$WQI = \frac{1}{n}\sum_{i=1}^{n} q_i$	q_i: rating scale for each parameter
Weighted arithmetic average	$WQI = \frac{1}{n}\sum_{i=1}^{n} q_i w_i$	w_i: weight unit of each parameter or weighting factor for the parameter i
Unweighted geometric average	$WQI = \left[\prod_{i=1}^{n} q_i\right]^{\frac{1}{n}}$	n: total number of parameters
Weighted geometric average	$WQI = \left[\prod_{i=1}^{n} q_i\right]^{w_i}$	
Minimum subscript	$WQI = Min(q_1, q_2, \ldots.. q_n)$	*Min*: minimum number of parameters
Maximum subscript	$WQI = Max(q_1, q_2, \ldots.. q_n)$	*Max*: maximum number of parameters
Modified unweighted average	$WQI = \frac{1}{100}\left[\frac{1}{n}\sum_{i=1}^{n} q_i\right]^2$	
Modified weighted average	$WQI = \frac{1}{100}\left[\frac{1}{n}\sum_{i=1}^{n} q_i w_i\right]^2$	

Source Samboni et al. (2007); Arriaga-Gaona et al. (2009); Al-Kareem and Al-Kizwini (2022); Abdo et al. (2021)

2.5 General Procedure for Formulating a WQI

According to Lino (2009), the following steps must be followed to formulate WQI:

(a) *Selection of parameters to be evaluated:* Ranging from 2 to *n* variables, the selection depends on the existing information, the time, the location, the expert professional, the importance of the source, and the required quality standard.
(b) *The values for each variable or subscript are determined:* Its purpose is to transform variables from a dimensional scale to a dimensionless scale to allow their aggregation. This is the most important stage.
(c) *The index is determined by the aggregation of subindices:* At this stage, the final information must be unified; the integration of subscripts determines WQI, which can be given by means of mathematical aggregation formulas that commonly correspond to an average function. According to Van Helmond and Breukel R. (1997), equations for the aggregation of subscripts are detailed in Table 4.

According to Fernández and Solano (2004), several methods can be used:

- *Parameter under mathematical formulation:* In order to convert the parameter values according to various scales with which the parameter values retain their original units.
- *Nominal or numerical value:* In this, the comparison is made between the value of the parameter and the standard value.
- *Parameter in decimal number, diagrams, or calibration tables:* A diagram is developed for each of the selected parameters indicating the correlation between the said parameter and the value on the quality scale, whose values could be between 0 and 100 or 0 and 1.0. According to Del Río (1986), the constructed curves are based on the following four methods:
- *Curves based on mathematical equations:* Mathematical expressions are used to develop the quality curve of each parameter, or the mathematical expression is established from the curves generated.
- *Method based on own experience:* It is very subjective because not only are there great differences in criteria between different authors but differences are also observed between the curves developed by the same author for different parameters.
- *The Delphi method:* For its construction, the average of the opinions of different experts is used.
- *Curves based on regulations:* Curves are generated from the values of the parameters collected in different regulations. The main objective of this method is to seek objectivity and acceptance by experts.

2.6 Advantages and Limitations of Water Quality Indices

As all the processes done manually are susceptible to errors, some advantages and limitations in the use of WQIs are presented below (Madsen et al. 2018; Zandbergen and Hall 1998; Sánchez et al. 2010).

Advantages

- Useful in evaluating the quality of water for different uses.
- It transforms a variety of environmental indicators into an easy-to-communicate system so that any user can easily understand it.
- It is a simple, straightforward, and validated method to express the importance of the data.
- Shows both spatial and temporal variation in the quality of a water source.
- Identifies trends in water quality and problem areas for the establishment of standards.
- Allows prioritising water quality management.

Limitations

- The information provided is not complete, as it cannot assess all the risks present in water.
- An index is limited to temporary spaces and can give wrong readings at a specific place and time.
- Some concepts are not applicable universally because of different environmental conditions that occur between two sources.
- Some scientists reject the methodology applied for the evaluation of indices, hence it could lose its credibility.
- An index may indicate that the water source is not suitable for one particular use but for another or others.

2.7 Some Indices of Contamination

2.7.1 WQI-NSF

Several researchers supported by the NSF of the United States proposed an index based on the structure of the Horton index, known as the NSF water quality index (WQI-NSF). Its classification is aimed at water intended to be used as a source of catchment for human consumption (Castro et al. 2014). The ICA is used to express water quality through a single value by aggregating the values of several physicochemical parameters (Khan et al. 2021). To establish this index, it is necessary to evaluate nine parameters, and for each parameter, a standard value is recognised, which are shown in Table 5 (Karunanidhi et al. 2021).

Table 5 NSF water quality indices

Parameter	WQI weight
Dissolved oxygen	0.17
Feacal coliforms (FCs)	0.15
pH	0.12
BOD_5	0.10
Nitrates	0.10
Total phosphates	0.10
ΔT (°C) equilibrium	0.10
Turbidity	0.08
Total solids (TS)	0.08

BOD5: biological oxygen demand in 5 days
Source Castro et al. (2014); Tunas (2014); Ramírez et al. (1997); Jiménez and Vélez (2006); Madsen et al. (2018); Torres Vega (2009); Lino et al. (2009); De et al. (2013); Suparno et al. (2021); Sheikh Khozani et al. (2022)

Table 6 Calculation sheet for the ICA

Parameter	Units	Sub w_i	Total
Dissolved oxygen	Saturation (%)	0.17	
Fecal coliforms (FCs)	NMP/100 ml	0.15	
pH	pH units	0.12	
DBO_5	mg/lt	0.10	
Nitrates	mg/lt	0.10	
Total phosphates	mg/lt	0.10	
ΔT equilibrium	°C	0.10	
Turbidity	FAU	0.08	
Total solids (TS)	mg/lt	0.08	
ICA value			

Source Kachroud et al. (2019); Torres Vega (2009)

The equations used to calculate this index are as follows:

$$ICA_a = \sum_{i=1}^{9} (sub_i w_i)$$

$$ICA_m = \prod_{i=1}^{9} \left(sub_i^{w_i}\right)$$

where w_i: relative weights assigned to each parameter (Sub_i).

Calculation of subscripts: These are made by using figures and tables. See Fig. 1(a)–(i) and Table 6.

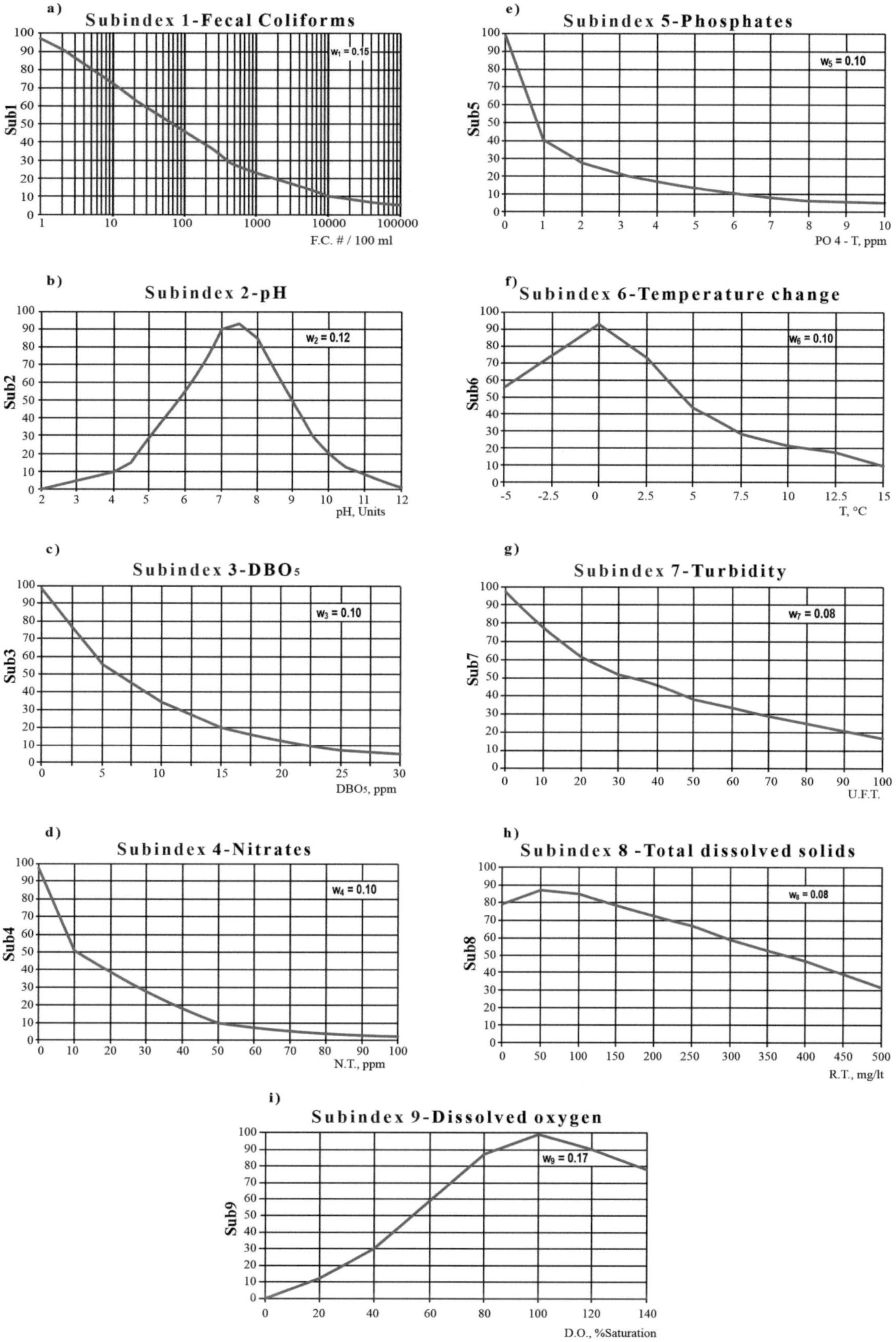
a)
Subindex 1-Fecal Coliforms
w1 = 0.15
Sub1
F.C. # / 100 ml
e)
Subindex 5-Phosphates
w5 = 0.10
Sub5
PO 4 - T, ppm
b)
Subindex 2-pH
w2 = 0.12
Sub2
pH, Units
f)
Subindex 6-Temperature change
w6 = 0.10
Sub6
T, °C
c)
Subindex 3-DBO5
w3 = 0.10
Sub3
DBO5, ppm
g)
Subindex 7-Turbidity
w7 = 0.08
Sub7
U.F.T.
d)
Subindex 4-Nitrates
w4 = 0.10
Sub4
N.T., ppm
h)
Subindex 8 -Total dissolved solids
w8 = 0.08
Sub8
R.T., mg/lt
i)
Subindex 9-Dissolved oxygen
w9 = 0.17
Sub9
D.O., %Saturation

◂**Fig. 1** Assessment of water quality. (Samboni et al. 2007; Fernandez and Solano 2007). *Subindex 1, Fecal coliforms:* If the value of fecal coliforms is more than 100,000 Bact/100 mL, then ($Sub\ w_i$) = 3. If the value of fecal coliforms is less than 100,000 Bact/100 mL, then look for the value on the '*X*' axis and proceed to interpolate the value on the '*Y*' axis (Fig. 1a). *Subindex 2, pH:* For $pH < 2$ units, the subscript two = 2. For $pH > 10$ units, the subscript two = 3 (Fig. 1b). *Subindex 3, demand of* chemical oxygen in 5 days *(DBO_5):* For $DBO_5 > 30$ mg/*lt*, the subindex 3 = 2 (Fig. 1c). *Subindex 4, nitrates:* For nitrate values >100 mg/*lt*, the subindex 4 = 2 (Fig. 1d). *Subindex 5: phosphates:* For phosphate values >10 mg/*lt*, the subindex 5 = 5 (Fig. 1e). *Subindex 6, temperature change:* Calculate temperature difference between the environment and the sample. If the difference is more than 15 °C, then the subindex 6 = 9 (Fig. 1f). *Subindex 7, turbidity:* For turbidity values >100 FAU, the subindex 7 = 5 (Fig. 1g). *Subindex 8, total dissolved solids* (TDS)*:* For TDS values >500 mg/*lt*, the subindex 8 = 3 (Fig. 1h). *Subindex 9, dissolved oxygen:* For dissolved oxygen saturation >140%, the subindex 9 = 47. The data obtained are integrated in the table to make the corresponding weighting (Fig. 1i)

2.7.2 WQI-Dinuis

Dinuis proposed a WQI comprising nine physicochemical and two microbiological variables. For its part, the WQI-INSF index comprises eight physicochemical variables and a biotic element, 'fecal coliforms'. It considers five uses of the source: human consumption, agriculture, fishing and aquatic life, industrial, and recreation. The subscripts of the Dinius index were developed from a review of the published scientific literature.

2.7.3 Universal Water Quality Index (UWQI) from Europe

The UWQI takes into account a total of the following 12 parameters: total coliforms, cadmium, cyanide, mercury, selenium, arsenic, fluorides, nitrates, dissolved oxygen, *pH*, *BOD*, and total phosphorus (Chávez 2010).

2.7.4 Simplified Index of Water Quality (WQSA)

The WQSA uses the product of temperature by the sum of the values obtained through equations that transform the concentrations of *COD*, total suspended solids (*TSS*), *DO,* and conductivity into a dimensionless number, just as they do other WQIs (Torres Vega 2009).

2.7.5 IAP

The IAP index is the resulting product of two indices (IQA and ISTO), and is made up of different aggregation functions; IQA, being an adaptation of WQI-NSF, uses an equation based on the weighted average, and in the case of ISTO, the equation used is the product of the weighting of two most critical minimum subindices of the group of toxic substances (*ST*) by the weighting obtained through the arithmetic mean of the subindices of the group of organoleptic substances (*SO*) (Torres Vega 2009).

2.7.6 ICAUCA

This index was based on the behaviour of water quality of the Cauca River and on the review of different WQIs developed globally from which the parameters, subindices, and equations to be considered in it were defined. This study was prepared by the CVC and the Universidad del Valle in 2004. It involves the following 10 parameters: *pH*, *OD*, colour, turbidity, BOD_5, total nitrogen, total phosphorus, *TS*, *TSS,* and FCs (Castro et al. 2014). To determine the ICAUCA calculation equation, two basic techniques, called additive and multiplicative techniques, were studied. In additive techniques, the WQI_a is obtained from the sum of the resulting products between subscripts and weights or weights of each variable, while in the multiplicative techniques, the WQI_m is calculated by multiplying the subscripts, placing the weights of the variable string as exponents in each one of them (Fernández et al. 2004a, b).

2.8 Equations Used to Calculate WQIs

Table 7 summarises the equations used to calculate WQIs by different authors.

$$F_1 = \left(\frac{failed\ parameters}{total\ parameters}\right) * 100$$

$$F_2 = \left(\frac{failed\ tests}{total\ tests}\right) * 100$$

$$excursion_1 = \left(\frac{failed\ value_1}{normed\ value_1}\right) - 1$$

In case the calculated value is not less than the normalised value, the following form is calculated:

$$excursion_1 = \left(\frac{normed\ value_1}{failed\ value_1}\right) - 1$$

Next, the normalised sum of excursions (*nse*) is calculated as follows:

$$nse = \frac{\sum_{i=1}^{n} excursion_1}{total\ tests}$$

Finally, the is calculated as follows:

$$F_3 = \frac{nse}{0.01nse + 0.01}$$

Table 7 Calculation equations used for the determination of WQIs

Group	Index	Equation	Meaning of terms
1	WQI-NSF (EU) WQI-Dinius (EU) IQA CETESB (Brazil) WQI-Rojas (Colombia) ICAUCA (Colombia)	$ICA_m = \prod_{i=1}^{n} (I_i^{w_i})$	i: corresponds to each of the chosen quality parameters. I_i: subindex of the parameter and w_i: weight or percentage assigned to each parameter i
2	CCME-WQI (Canada) DWQI (EU)	$ICA = 100 - \frac{\sqrt{[F_1^2+F_2^2+F_3^2]}}{1.732}$	$F1$: percentage of parameters exceeding the norm $F2$: percentage of individual tests of each parameter that exceeds the standard $F3$: magnitude by which each parameter that is not met exceeds the standard
3	UWQI (Europe)	$UWQI = \sum_{i=1}^{n} W_i I_i$	Weighted arithmetic average W_i: weight or percentage assigned to the ith parameter I_i: subindex of ith parameter
4	WQSA (Spain)	$ISQA = \mathrm{T}\,(COD + SS + OD + Cond)$	T: Temperature COD: chemical oxygen demand
			OD: dissolved oxygen Cond: conductivity SS: suspended solids As of 2003, the WQSA began to be calculated by replacing COD with total organic carbon (TOC in mg/L)
5	IAP (Brazil)	$IAP = ISTO * IQACETESB$ $ISTO = ST * SO$ $ST = Min - 1(q_1, q_2, \ldots q_n) *$ $Max - 2(q_1, q_2, \ldots q_n)$ $SO = Arithmetic\ average(q_a, q_b, q_n)$	IQA: water quality index adapted from the ICA-NSF for the conditions of Brazil $ISTO$: index of toxic and organoleptic substances ST: weighting of the two most critical minimum subindices of the group of toxic substances SO: weighting obtained through the arithmetic mean of the subindices of the group of organoleptic substances

Source Khatri et al. (2020); Valcarcel et al. (2009); Torres Vega (2009);Islam and Mostafa (2022)

2.9 Physicochemical and Microbiological Parameters Used by Different *WQIs*

The WQIs are varied and involve different parameters. Table 8 consolidates the indices used in some countries with their respective physicochemical and microbiological parameters used to obtain them.

3 Water Contamination Indices in Colombia

At the national level, Ramírez et al. (1997) based on the physicochemical information of limnology studies related to the Colombian oil industry developed pollution indices for determining the degree of water pollution by grouping physicochemical variables. This topic began its study in 1997 from the monitoring carried out by oil companies. In its development, the existing Colombian and international regulations for various uses of water were taken into account, with the aim of achieving a generalisation in its application and its use in different situations. The methodological procedure began with the assignment of values between 0.0 and 1.0 to the scale of variables. Then the equation that relates the value of the variable and its incidence in contamination was selected. Subsequently, the linear regression analysis was made by the method of ordinary least squares to the relationship between the index and its parameter, and finally the adjustments were made to the given equation (Ramírez 2004; Cañas 2006).

3.1 Mineralization Contamination Index (MCI)

Dissolved solids, hardness by calcium and magnesium cations, and alkalinity by anions integrate conductivity, hardness and alkalinity. Its expression is as follows:

$$MCI = \left(I_{Conductivity} + I_{Hardness} + I_{Alkalinity}\right)$$

$$I_{Conductivity} = Log_{10} I_{Conductivity} = 3.26 + 1.34\,Log_{10}\ Conductivity\ \left({}^{\mu S}/_{cm}\right)$$

$$I_{Conductivity} = 10^{Log I_{Conductivity}}$$

For conductivity greater than 270, conductivity index = 1

$$I_{Hardness} = Log_{10} I_{Hardness} = -9.09 + 4.40\ Log_{10} Hardness\left({}^{mg}/_{lt}\right)$$

$$I_{Hardness} = 10^{\log I_{Hardness}}$$

For hardness greater 110 mg/ltL, conductivity index = 1

For hardness less than 30 mg/ltL, conductivity index = 0

$$I_{Alkalinity} = -0.25 + 0.005 Alkalinity\left({}^{mg}/_{lt}\right)$$

Table 8 Physicochemical and microbiological parameters used by different WQIs

Country	United states		UNEP-GEMS		European Union	Spain	Brazil		Colombia	
Index	WQI-NSF	WQI-dinius	DWQI		UWQ 12,007	WQSA 1982	IAP		ICA-Rojas 1991	ICAUCA 2004
			HWQI 2007	AWQI 2007			IQA 1975	ISTO 2002		
Parameters										
DO	X	X			X	X	X		X	X
pH	X			X	X		X		X	X
COD	X	X			X					
Nitrates	X	X	X				X		X	X
Fecal coliforms (FCs)	X	X					X			
Temperature	X	X				X	X		X	X
Turbidity	X						X		X	X
Total dissolved solids (TDS)	X									
Total phosphorus					X		X			X
Cadmium			X		X			X		
Mercury			X		X			X		
Conductivity		X				X				
Suspended solids						X				X
Colour		X								X
Lead			X					X		
Total chrome			X					X		
Arsenic			X		X	X	X		X	X
Fluorides			X							
Manganese			X					X		
Zinc				X				X		
Total coliforms		X			X					
COD						X				
Alkalinity		X								
Hardness		X								
Nitrites			X							
Ammonia				X						
Phosphates	X									
Sodium				X						
Sulphates				X						
Iron				X						
Cooper			X							
Borom			X							
Nickel								X		
Cyanide					X					

(continued)

Table 8 (continued)

Country	United states		UNEP-GEMS		European Union	Spain	Brazil		Colombia	
Index	WQI-NSF	WQI-dinius	DWQI		UWQ 12,007	WQSA 1982	IAP		ICA-Rojas 1991	ICAUCA 2004
			HWQI 2007	AWQI 2007			IQA 1975	ISTO 2002		
Selenium					X					
PFTHM (trihalomethanes formation potential)								X		
Dissolved aluminum								X		
Dissolved copper								X		
Dissolved iron								X		
Total organic carbon (TOC)						X				
Total parameters	9	12	18	18	12	5	20	20	6	10

Source Kachroud et al. (2019); Fernández and Ramírez (2003)

For alkalinity greater than 250 mg/*lLt*, conductivity index = 1

For hardness less than 50 mg/*lt*L, conductivity index = 0.

3.2 Index of Contamination by Suspended Solids (ICOSUS)

Determined by concentration of suspended solids

$$ICOSUS = -0.02 + 0.0003\, Suspended\ Solids({}^{mg}/_{lt})$$

For suspended solids > 340 mg/lt, $ICOSUS = 1$,

For suspended solids < 10 mg/lt, $ICOSUS = 0$.

Trophic contamination index (TCI) is calculated based on the concentration of total phosphorus.

Oligotrophy: < 0.01,	Eutrophy: 0.02–1.00,
Mesotrophy: 0.01–0.02,	Hipereutrophy: > 1.00.

3.3 Temperature Contamination Index (TEMPCOI)

This index is formulated for the evaluation of discharges and the receiving body. Its mathematical expression is:

$$TEMPCOI = -0.49 + 127 Log(\text{Pouring Temperature} - \text{Receiver Course Temperature})$$

If the temperature difference is less than 2.5 °C (275.5°K), $TEMPCOI = 0$.

If the temperature difference is greater than 15 °C (288.5°K), $TEMPCOI = 1$.

3.4 PH Contamination Index (*pHCI*)

$$pHCI = \frac{e^{31.08+3.45\,PH}}{1 + e^{-31.08+3.45\,PH}}$$

Regarding established ranges, $pHCI$ values are summarised in Table 9.

Table 9 Water contamination indices (WCI)

WCI	*Contamination grade*	*Colour scale*
0–0.2	None	
>0.2–0.4	Low	
>0.4–0.6	Half	
>0.6–0.8	High	
>0.8–1.0	Very high	

Source Ramírez 2004; Chavarro and Gelvez 2016

3.5 Contamination Index by Organic Matter (*CIOM*)

This index is made up of biochemical oxygen demand (*BOD*), ammoniacal nitrogen, phosphorous nitrites, *COD*, total coliforms, and oxygen saturation.

$$CIOM = \frac{1}{3}\left(I_{COD} + I_{Coliforms} + I_{Oxygen}\right)$$

$$I_{BOD} = -0.05 + 0.70\, Log_{10} BOD\left({}^{mg}/_{lt}\right)$$

$$BOD > 30(\text{mg/L}) = 1$$

$$BOD > 2(\text{mg/L}) = 0$$

$$I_{Total\ Coliforms} = 1.14 + 0.56\, Log_{10} BOD\left({}^{mg}/_{lt}\right)$$

$$I_{Total\ Coliforms} > 20.000(\text{NMP}/100\,\text{mL}) = 1$$

$$I_{Total\ Coliforms} < 500(\text{NMP}/100\,\text{mL}) = 0$$

$$I_{Oxygen\%} = 1 - 0.01\ Oxygen\% - 1$$

Oxygen (%) > 100% has an oxygen index of 0.

For lentic systems with eutrophication and saturation percentages greater than 100%, it is suggested to replace the expression with the following:

$$I_{Oxygen\%} = 0.01\ Oxygen\% - 1$$

3.6 Quality Risk Index (*QRI*)

The water quality risk index measures the degree of risk of occurrence of diseases related to non-compliance with the physical, chemical and microbiological characteristics of water for human consumption once it has been subjected to different treatment processes that guarantee its potability (Torres Vega 2009).

4 The Most Used Water Quality Indices Worldwide

In a recent study carried out from 1960 to 2020 in relation to surface WQIs, it was determined that these indices were used maximum in the United States, Canada, Indonesia, and Malaysia, as seen in Fig. 2a (Uddin et al. 2021). On the other hand, the countries where indices have been studied most and/or used to determine the quality of water in their water sources are China, the United States, India, Brazil, Italy, Spain, Iran, the United Kingdom, Canada, Germany, etc., as seen in Fig. 2b.

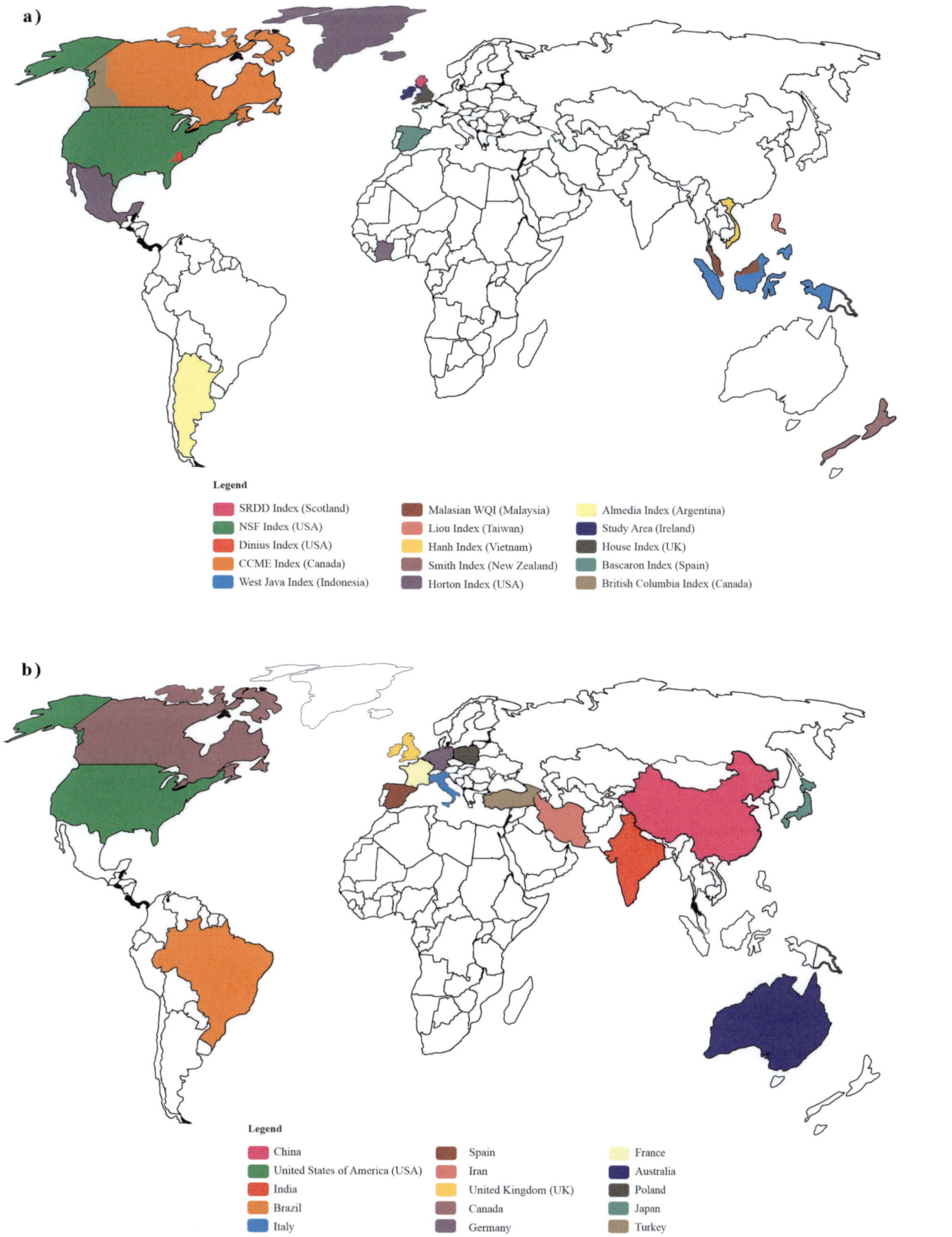

Fig. 2 **a** Countries with the most used water quality indices (1960–2020; Uddin et al. 2021); **b** countries where water quality indices are studied most

5 Conclusion

Each WQI aims to reduce different parameters to be taken into account when evaluating a water source to a single variable, estimating, of course, the regulations applicable to each region or country as well as technological advances. Because water quality varies both spatially and temporally in relation to its uses, this aspect must also be taken into account along with the variables used in measuring the WQI. Thus, the number of variables to analyse or measure should neither be limited in that they are not decisive when establishing the quality of the source nor should they be too many that it becomes impossible to evaluate the index.

Throughout the recent history of WQIs, various forms have been used to add subindices; among them are arithmetic average, weighted arithmetic average, weighted geometric average, and fuzzy logic. All these methods require that there is no subjectivity in the data; on the other hand, the ideal would be that WQIs must be established according to the region or country in order to involve the environmental characteristics of their surroundings and also to do it in a comprehensive manner that includes all or the vast majority of variables to analyse.

Currently, WQIs are valid indices for the evaluation of physical, chemical and biological parameters. WQIs use a variety of parameters which can later be added to a single value, being understandable or interpreted by any person, whether an expert or from the community in general. This allows quick action to be taken at local, national, and international levels and therefore direct political-legal actions towards the preservation of water resources.

Parameters that are mainly used in the evaluation of water sources include variables, such as dissolved oxygen; potential of hydrogen, which can cause changes in the composition of the fauna and flora of water sources and also reduces the toxicity because of certain substances; and demand of chemical oxygen in 5 days, associated with the quality of water for the fish habitat and the nitrogen that in high concentrations give rise to eutrophication or the presence of some algae that can reduce the amount of oxygen in the bodies of water.

Nowadays, there is no universal consensus regarding the use of necessary subindices for the calculation of the WQI, which in a certain way becomes a costly step; but this variation of criteria is more than understandable, given the variability of ecosystems in the world. The dynamics that the same water source still varies both spatially and temporally depending on its location, for example, the area of the globe in which it is found, the stage of the source, whether it is in its upper part (river child), middle part (river young) or lower part (old river), will have very different characteristics. In addition to the above, the population increase and the anthropic action that this may generate in the water source must be taken into account, either directly or indirectly.

One of the institutions that has carried out a good number of investigations in relation to WQIs is the NSF, and this, in turn, is the most used and modified methodology in various countries around the world, including Colombia. In Colombia, complementary indices have been developed in the department of Valle del Cauca as well as in regions where their industrial production is based on hydrocarbons or the extraction of precious metals. It should be clarified that the NSF index only contemplates nine parameters that do not include variables or aspects related to the ICO, ICOMO, or ICOTRO contamination indices. For this reason, it is necessary to take into account, in addition to the NSF, the contamination indices and thus obtain a better assessment of the quality of a water source. These contamination indices are more relevant to the extent that the water source to be analysed is located in industrial or mining areas because of the inputs and waste generated in the production of these goods and services. In that order of ideas, countries like Colombia should implement actions aimed at permanent monitoring of the quality of surface water using the WQIs' water quality indices and ICO contamination. This allows corrective actions to be taken and prevent the deterioration of rivers and other bodies of water, both surface and underground.

Therefore, before using any methodology to establish the WQI, it is necessary to identify the methodological bases taken into account and subsequently define its applicability or not in the specific region, although the best option would be the elaboration or adaptation of the index, since, in addition to local regulations, the area of influence of the water source would be taken into account. Finally, the WQIs can be determined with methodologies that combine or include geographic information system tools, satellites, artificial intelligence, or any other type of instrument that facilitates their evaluation and calculation. This will depend a lot on the area of influence, the use that the source will have, and the technology available.

References

Abdel-Satar, A.M., Ali, M.H. and Goher, M.E. (2017). Indices of water quality and metal pollution of Nile river, Egypt. *Egyptian Journal of Aquatic Research*, **43(1):** 21–29. https://doi.org/10.1016/j.ejar.2016.12.006

Abdo, M., Ahmed, H., Helal, M., Fekry, M. and Abdelhamid, A. (2021). Water quality index and environmental assessment of Rosetta branch aquatic system, Nile river, Egypt. *Egyptian Journal of Chemistry*, **0(0):** 0–0. https://doi.org/10.21608/ejchem.2021.92605.4405

Abed, B.S., Farhan, A.A.R., Ismail, A.H. and Al Aani, S. (2022). Water quality index toward a reliable assessment for water supply uses: A novel approach. *International Journal of Environmental Science and Technology*, **19(4):** 2885–2898. https://doi.org/10.1007/s13762-021-03338-7

Agrama, A.A. (2012). Assessment of surface water quality in Egypt. *Journal of Applied Sciences Research*, **8(4):** 1944–1951.

Agua, C.D.E.L. and Villeta, R. (2010). *Cap6 Capítulo 6.*

Al-Kareem, S.A. and Al-Kizwini, R.S. (2022). Statistical analysis for water quality index for Shatt-Al-Hilla river in Babel city. *Water Practice and Technology*, **17(2):** 567–586. https://doi.org/10.2166/wpt.2022.004

Álvarez, J.P.A., Panta, J.E.R., Ayala, C.R. and Acosta, E.H. (2008). Calidad integral del agua superficial en la cuenca hidrológicadel río Amajac. *Informacion Tecnologica*, **19(6):** 21–32. https://doi.org/10.1612/inf.tecnol.3975it.07

Arriaga-Gaona, M.L., Hernández-Limón, L., Sandoval-Reyes, F., Vera-Lara, J., Vargas-Moreno, E. and García-Rojas, L. (2009). Monitoreo de la calidad del agua del río Tecolutla desde Coyutla hasta Gutierrez Zamora, Veracruz. *Revista Latinoamericana de Recursos Naturales*, **5(2):** 141–147.

Auge, M. (2007). Agua Fuente de Vida (Water Source of Life). Salud Mundial, 31. WHO, Geneva, Switzerland.

Bajaña, L.S., Martínez-Fresneda, M., Navarro, J.C., de S. Celente, G. and Lobo, E.A. (2022). Water quality index (WQI) calibration in the Paute river hydrographical basin, south inter-Andean region of Ecuador, based on the environmental agreement n° 097-A. *Sustainable Water Resources Management*, **8(1):** 1–12. https://doi.org/10.1007/s40899-021-00590-0

Boyacioglu, H. (2007). Development of a water quality index based on a European classification scheme. *Water SA*, **33(1):** 101–106. https://doi.org/10.4314/wsa.v33i1.47882

Bustamante Toro, C.A., Monsalve Durango, E.A. and García Reinoso, P.L. (2008). Análisis de la calidad del agua en la cuenca media del río Quindío con base en índices físicos, químicos y biológicos (Analysis of the water quality at the intermediate river basin of the Quindío river on the basis of physical, chemical and biological inde). *Revista de Investigaciones Universidad Del Quindío*, **18:** 22–31.

Caho-Rodríguez, C.A. and López-Barrera, E.A. (2017). Determinación del Índice de Calidad de Agua para el sector occidental del humedal Torca-Guaymaral empleando las metodologías UWQI y CWQI. *Producción + Limpia*, **12(2):** 35–49. https://doi.org/10.22507/pml.v12n2a3

Cañas, J. (2006). Determinación y Evaluación de Índices de Contaminación (ICOs) en Cuerpos de Agua. Universidad Militar Nueva Granada, Bogotá, Cundinamarca, Colombia. Available at: http://repository.unimilitar.edu.co/bitstream/10654/10901/1/articulo final.pdf

Castro, M., Almeida, J., Ferrer, J. and Diaz, D. (2014). Indicadores de la calidad del agua: evolución y tendencias a nivel global. *Ingeniería Solidaria*, **10(17):** 111–124. https://doi.org/10.16925/in.v9i17.811

Chavarro, A.G. and Gelvez Bernal, E.J. (2016). Caracterización de la calidad de las aguas de la quebrada Fucha utilizando los índices de contaminación ICO con respecto a la precipitación y usos del suelo. *Revista Mutis*, **6(2):** 19–31. https://doi.org/10.21789/22561498.1148

Chávez L. (2010). Evaluación espacial y temporal del índice de calidad del agua del río Cazones en Coatzintla [Tesis Pregrado, Universidad Veracruzana]. https://www.uv.mx/pozarica/mca/files/2019/05/G02_Luz-Nalleli-Chavez-Martinez.pdf

Choque-Quispe, D., Froehner, S., Palomino-Rincón, H., Peralta-Guevara, D.E., Barboza-Palomino, G.I., Kari-Ferro, A., Zamalloa-Puma, L.M., Mojo-Quisani, A., Barboza-Palomino, E.E., Zamalloa-Puma, M.M.; et al. (2022). Proposal of a Water-Quality Index for High Andean Basins: Application to the Chumbao River, Andahuaylas, Peru. *Water (Switzerland)*, **14(4)**. https://doi.org/10.3390/w14040654

Corporación Autónoma Regional del Valle del Cauca (CVC) & Universidad del Valle. (2004). Proyecto de modelación del río Cauca - PMC-Fase II., Volumen X, Estudio de la calidad del agua del río Cauca y sus principales tributarios mediante la aplicación de índices de calidad y contaminación tramo Salvajina-la Virginia. Convenio Interadministrativo 0168 de Noviembre 27 de 2002.

Debels, P., Figueroa, R., Urrutia, R., Barra, R., & Niell, X. (2005). Evaluation of water quality in the Chillán River (Central Chile) using physicochemical parameters and a modified Water Quality Index. *Environmental Monitoring and Assessment*, **110(1–3)**, 301–322. https://doi.org/10.1007/s10661-005-8064-1

Dos Santos, D.A., Molineri, C., Reynaga, M.C. and Basualdo, C. (2011). Which index is the best to assess stream health? *Ecological Indicators*, **11(2):** 582–589. https://doi.org/10.1016/j.ecolind.2010.08.004

Dutta, N., Thakur, B.K., Nurujjaman, M., Debnath, K. and Bal, D.P. (2022). An assessment of the water quality index (WQI) of drinking water in the Eastern Himalayas of South Sikkim, India. *Groundwater for Sustainable Development*, **17(Feb):** 100735. https://doi.org/10.1016/j.gsd.2022.100735

Espejo, L.M. (2010). Cálculo de índices de calidad de aguas superficiales y análisis de la red de monitoreo en las cuencas de Huasco, Elqui, Limarí y Choapa [Tesis de Pregrado, Universidad de La Serena]. https://es.slideshare.net/MitziLaurin/indice-de-calidadaguassuperficialestesis

Fernández, N. Ramírez, A. and Solano, F. (2003). Índices fisicoquímicos de calidad del agua un estudio comparativo. *Conferencia Internacional Usos Múltiples Del Agua: Para La Vida y El Desarrollo Sostenible*, **81(3):** 211–219.

Fernández, N., Ramírez, A. and Solano, F. (2004). Physico-chemical water quality indices – A comparative review. *Bistua: Revista de La Facultad de Ciencias Básicas*, **2(1):** 19–30. https://doi.org/10.24054/01204211.v1.n1.2004.9

Fernández, N., Ramos, G. and Solano, F. (2004). Icatest V 1.0 Una herramienta Informática para el análisis y valoración de la calidad del Agua. *Bistua Universidad de Pamplona*, *2*, 88–97. https://www.unipamplona.edu.co/unipamplona/portalIG/home_10/recursos/general/pag_contenido/publicaciones/bistua_revista_ciencias_basica/2004/11082010/rev_bis_vol2_num2_art12.pdf

Fernandez, N.J. and Solano, F. (2007). Capítulo III: índices de calidad (ICAs) y de contaminación (ICOs) del agua de importancia mundial, 76.

Fernández N. and Solano F. (2005). Índices de calidad (ICAs) y de contaminación (ICOs) del agua de importancia mundial. In *Índices de Calidad y de Contaminación del Agua* (pp. 1—78). Universidad de Pamplona. https://www.unipamplona.edu.co/unipamplona/portalIG/home_10/recursos/general/pag_contenido/libros/06082010/icatest_capitulo3.pdf

Fontalvo Julio, F.A. and Tamaris-Turizo, C.E. (2018). Water quality in the lower section of Córdoba River (Magdalena, Colombia), using the ICA-NSF. *Intropica*, **2:** 101–112. https://doi.org/10.21676/23897864.2510

Giacomett, J. and Bersosa, F. (2006). Macroinvertebrados acuáticos y su importancia como bioindicadores de calidad del agua en el río Alambi. *Serie Zoológica*, **2(Appendix 1):** 17–32.

Giao, N.T., Nhien, H.T.H., Anh, P.K. and Van Ni, D. (2021). Classification of water quality in low-lying area in Vietnamese Mekong delta using set pair analysis method and Vietnamese water quality index. *Environmental Monitoring and Assessment*, **193(6):** 1–16. https://doi.org/10.1007/s10661-021-09102-1

Gómez Delgado, M. (1995). El estudio de los residuos: Definiciones, tipologías, gestión y tratamiento. *Serie Geográfica*, **5:** 21–42.

González V., Caicedo O. and Aguirre N. (2012). Aplication of water quality indeces NSF, DINIUS and BMWP to Ayurá Creek, Antioquia, Colombia. *Revista Gestión y Ambiente*, **16(1):** 97–107. https://www.redalyc.org/pdf/1694/169427489003.pdf

Islam, M.S. and Mostafa, M.G. (2022). Development of an integrated irrigation water quality index (IIWQIndex) model. *Water Supply*, **22(2):** 2322–2337. https://doi.org/10.2166/ws.2021.378

Jiménez, M.A. and Vélez, M.V. (2006). Analisis comparativo de indicadores de la calidad de agua superficial. *Avances En Recursos Hidráulicos*, **14:** 53–69.

Kachroud, M., Trolard, F., Kefi, M., Jebari, S. and Bourrié, G. (2019). Water quality indices: Challenges and application limits in the literature. *Water* (Switzerland), **11(2):** 1–26. https://doi.org/10.3390/w11020361

Karunanidhi, D., Aravinthasamy, P., Subramani, T. and Muthusankar, G. (2021). Revealing drinking water quality issues and possible health risks based on water quality index (WQI) method in the Shanmuganadhi river basin of South India. *Environmental Geochemistry and Health*, **43(2):** 931–948. https://doi.org/10.1007/s10653-020-00613-3

Khan, R., Saxena, A., Shukla, S., Sekar, S. and Goel, P. (2021). Effect of the COVID-19 lockdown on the water quality index of river Gomti, India, with potential hazard of faecal-oral transmission. *Environmental Science and Pollution Research*, **28(25):** 33021–33029. https://doi.org/10.1007/s11356-021-13096-1

Khatri, N., Tyagi, S., Rawtani, D., Tharmavaram, M. and Kamboj, R.D. (2020). Analysis and assessment of ground water quality in Satlasana Taluka, Mehsana district, Gujarat, India through application of water quality indices. *Groundwater for Sustainable Development*, **10(13):** 100321. https://doi.org/10.1016/j.gsd.2019.100321andr:xxx.

Kumar, A., Bojjagani, S., Maurya, A. and Kisku, G.C. (2022). Spatial distribution of physicochemical-bacteriological parametric quality and water quality index of Gomti river, India. *Environmental Monitoring and Assessment*, **194(3):** 1–17. https://doi.org/10.1007/s10661-022-09814-y

Lino, V., Nancy, A. and Daniel, F. (2009). El Índice de calidad de agua como herramienta para la gestión de los recursos hídricos. *Medio Ambiente y Desarrolo; Revista Eléctronica de La Agencia de Medio Ambiente*, **16:** 1–5. http://www.ciesas.edu.mx/cee_xalapa/Informedegestionareassustantivas/Proyectosinstitucionales/aha.pdf

Lopes, O.F., de Jesus, R.M., de Sousa, L.F., Rocha, F.A., da Silva, D.M.L., Amorim, A.F., da Silva, V.H.C. and Navoni, J.A. (2021). Comparison between water quality indices in watersheds of the Southern Bahia (Brazil) with different land use. *Environmental Science and Pollution Research*, **28(10):** 12944–12959. https://doi.org/10.1007/s11356-020-10941-7

Madsen, L.B., Lévêque, C., Omiste, J.J. and Miyagi, H. (2018). Chapter 11: Time-dependent restricted-active-space self-consistent-field theory for electron dynamics on the attosecond timescale. *RSC Theoretical and Computational Chemistry Series*, **2018(13):** 386–423. https://doi.org/10.1039/9781788012669-00386

Pascual Aguilar, J.A., Campo, J., Meneu, S.N., Gimeno-García, E. and Andreu, V. (2019). Analysis of existing water information for the applicability of water quality indices in the fluvial-littoral area of turia and Jucar rivers, valencia, Spain. *Applied Geography*, **111(Aug):** 102062. https://doi.org/10.1016/j.apgeog.2019.102062

Pérez, J.I., Nardini, A.G. and Galindo, A.A. (2018). Comparative analysis of water quality indices applied to the ranchería river, la guajira-Colombia. *Informacion Tecnologica*, **29(3):** 47–58. https://doi.org/10.4067/S0718-07642018000300047

Posada C., Vanegas R., Domínguez E. and Rivera H. (2008). Informe anual sobre el estado del medio ambiente y los recursos naturales renovables en Colombia. Estudio Nacional del agua, relaciones de demanda de agua y oferta hídrica. http://www.ideam.gov.co/documents/14691/125687919/Estudio+Nacional+del+Agua+relaciones+de+demanda+de+agua+y+oerta+h%C3%ADdrica+2008.pdf/035f7eb6-5c30-4237-8808-6ff62fbae6a0?version=1.0

Ramírez, C.A. (2004). Identificación de parametros criticos en el río Cauca y sus principales ríos tributarios. 680.

Ramírez, A., Restrepo, R. and Viña, G. (1997). Cuatro índices de contaminación para caracterización de aguas continentales. Formulaciones y aplicación. *CT y F – Ciencia, Tecnologia y Futuro*, **1(3):** 135–153.

Reinoso-Florez, G., Villa-Navarro, F.A., Esquivel, H.E., Garciamelo, J.E.Y. and Vejarano-Delgado, M. (2008). Biodiversidad Faunística y Florística de la Cuenca mayor del rio Saldaña (subcuenca Anamichú). Biodiversidad Regional Fase IV (p. 150). Grupo de Investigación en Zoología, Universidad del Tolima, Ibagué, Colombia.

Ríos-Tobón, S., Agudelo-Cadavid, R.M. and Gutiérrez-Builes, L.A. (2017). Patógenos e indicadores microbiológicos de calidad del agua para consumo humano. *Revista Facultad Nacional de Salud Pública*, **35(2):** 236–247. https://doi.org/10.17533/udea.rfnsp.v35n2a08

Samboni, R.N.E., Carvajal, E.Y. and Escobar, J.C. (2007). Revisión de parámetros fisicoquímicos como indicadores de calidad y contaminación del agua. *Revista de Ingeniería e Investigación*, **27(3):** 172–181.

Sánchez, F., García, M., Jaramillo, O. and Verdugo, N. (2010). Capítulo 3. Agua superficial. Caracterización y análisis de oferta. *Estudio Nacional Del Agua*, **2010:** 69.

Sheikh Khozani, Z., Iranmehr, M. and Wan Mohtar, W.H.M. (2022). Improving water quality index prediction for water resources management plans in Malaysia: Application of machine learning techniques. Geocarto International, **0(0):** 1–18. https://doi.org/10.1080/10106049.2022.2032388

Simbaña-Farinango, K., Romero-Estévez, D., Yánez-Jácome, G., Benavides, D. and Navarrete, H. (2019). Evaluación de la calidad del agua del río Pita (Ecuador), implicación para la conservación de la vida acuática y silvestre. *Infoanalítica*, **7(2):** 71–93. https://doi.org/10.26807/ia.v7i2.104

Sousa, W., Attayde, J., Rocha, E., & Eskinazi-Sant'Anna, E. (2008). The response of zooplankton assemblages to variations in the water quality of four man-made lakes in semi-arid northeastern Brazil. *Journal of Plankton Research*, **30(6)**, 699–708. https://doi.org/10.1093/plankt/fbn032

Suparno, A., Moni, B., Abdullah, S., Farris, M., Leong, I., Id, A., Kabir, M.S., Id, S.M.A., Sultana, F., Salehin, M., Mohammed, S., Islam, S., Cross, W. and Rahman M.A.. (2021, Sep). Psychological distress, fear and coping among Malaysians during the COVID-19 pandemic. *PLoS One*, **16(19):** 1–21. https://doi.org/10.1371/journal.pone.0257304

Tibebe, D., Zewge, F., Lemma, B. and Kassa, Y. (2022). Assessment of spatio-temporal variations of selected water quality parameters of Lake Ziway, Ethiopia using multivariate techniques. *BMC Chemistry*, **16(1):** 1–18. https://doi.org/10.1186/s13065-022-00806-0

Torres Vega, F. J. (2009). Desarrollo y aplicación de un índice de calidad de agua para ríos en Puerto Rico [Tesis de grado, Universidad de Puerto Rico]. http://prwreri.uprm.edu/publications/PR_2009_01.pdf

Tunas, L. (2014). Índice canadiense de calidad de las aguas para la cuenca del río Naranjo, provincia Las Tunas, Cuba. *Revista Ciencias Técnicas Agropecuarias*, **23(3):** 11–16.

Uddin, M.G., Nash, S. and Olbert, A.I. (2021). A review of water quality index models and their use for assessing surface water quality. *Ecological Indicators*, **122:** 107218. https://doi.org/10.1016/j.ecolind.2020.107218

Valcarcel, L.R., Alberro, N.M. and Frías, D. (2009). El Índice de calidad de agua como herramienta para la gestión de los recursos hídricos. *Medio Ambiente y Desarrolo; Revista Eléctronica de La Agencia de Medio Ambiente*, **16:** 1–5. http://www.ciesas.edu.mx/cee_xalapa/Informedegestionareassustantivas/Proyectosinstitucionales/aha.pdf

Yimmy Montoya, M., Acosta, Y. and Zuluaga, E. (2011). Evolución de la calidad del agua en el río negro y sus principales tributarios empleando como indicadores los índices ICA, el BMWP/Col y el ASPT. *Caldasia*, **33(1):** 193–210.

Zandbergen, P.A. and Hall, K.J. (1998). Analysis of the British Columbia water quality index for watershed managers: A case study of two small watersheds. *Water Quality Research Journal of Canada*, **33(4):** 519–549. https://doi.org/10.2166/wqrj.1998.030

Zapata M.V., Zárate C.M., Aguilera V.J., Zelaya A.A., Barrios S.E., Laupheimer G.S., Aravena C.G., Garrido I.C. and Taub E.T. (2006). Indicadores de calidad en la gestion del personal. *Revista Chilena de Radiologia*, **12(4):** 157–160. https://doi.org/10.4067/s0717-93082006000400004

Policy Development for Carbon Sequestration: A Case Study on Mining Aspects in Mexico

María Concepción Martínez-Rodríguez, Ana Laura Cervantes-Nájera, and Martín C. Vera Martínez

1 Introduction

Mining marked, since the early man's appearance and with the discovery of new minerals, the Stone, Copper, Bronze, Iron, and Coal Age, alluding to the primary material from which everyday objects were made at that time (Violatti 2014). These practices favoured the later development of ornamental aspect with precious materials and promoted the exchange of the products made with the metals found in the areas where the rock deposits containing these minerals were abundant. Mining in the Americas and largely in Mexico formally began with the Spanish conquest, as the mining importance by the indigenous people before this period needed to be more relevant in terms of progress (López and Ruvalcaba 2015).

Currently, mining is the extraction of minerals from the earth's crust for economic benefits (Concha 2017). In economic terms, this industry correctly contributes 11.5% of the world's gross domestic product (GDP). The total number of mines worldwide varies, and a precise definition of what is considered a mine would need to be established. In the case of small-scale and artisanal miners, there are more than 8,300 miners in China alone (Parejo and Parejo 2012). In Mexico, mining has significant importance for its economy. Mexico is a world power country in mineral production, being a major producer of silver, lead, zinc, salt, gypsum, gold, and copper, among others (Gaytán and Benita 2014).

Industry, in general, is one area of concern for achieving sustainable development, as the production methods are in line with the time when global warming consequences seemed distant. In this sense, mining preserves production and consumption patterns that have a serious impact on the environment, leaving aside biological diversity and natural heritage, as well as socially with the lack of social demands and human rights recognition, and economically with deficient industrial policies, and inexperience in the income and accountability allocation.

Mining has significant potential to align with the Sustainable Development Goals (SDGs), as they can guide mining as a social investment and can implement programmes that benefit human rights, inclusion, and good environmental practices (Vela Almeida et al. 2021). Also, mining is related to the environmental, economic, and social areas (Alegría et al. 2021) according to the following SDGs. In terms of the environment, mining can have negative consequences for the climate, and resources, such as land, flora, fauna, and water. In these cases, mining must respond with actions to SDG 6 (Clean Water and Sanitation), SDG 7 (Affordable and Clean Energy), SDG 13 (Climate Action), and SDG 15 (Life of Wild Ecosystems).

In terms of economic and social issues, there are possible positive scenarios, such as job opportunities, as well as negative situations that put the community's human rights at risk. Mining plays an important role in the local, regional, and national economy by generating job opportunities and fostering the development of infrastructure for communication, transport, water, and energy. SDGs that correspond to the mining sector through the generation of jobs and economic growth are as follows: SDG 1 (End Poverty), SDG 5 (Gender Equality), SDG 8 (Decent Work and Economic Growth), SDG 9 (Industry, Innovation and Infrastructure) and SDG 10 (Inequalities Reduction). SDG 16 (Peace, Justice, and Strong Institutions) refers to the mining sector's role in supporting peaceful societies, and ensuring laws through prevention and conflict resolution between local communities and companies.

M. C. Martínez-Rodríguez · A. L. Cervantes-Nájera (✉)
Centro Interdisciplinario de Investigaciones y Estudios Sobre Medio Ambiente y Desarrollo (CIIEMAD), Instituto Politécnico Nacional (IPN), Alcaldía Gustavo A. Madero, Mexico City, Mexico
e-mail: cer.lau.ana.18@gmail.com

M. C. Vera Martínez
Parque Internacional Industrial Tijuana, Universidad Autónoma de Baja California, Universidad 14418, Tijuana, Baja California, Mexico

J. Marrugo-Negrete et al. (eds.), *Remediation and Carbon Capture in Mining Sector*, https://doi.org/10.1007/978-3-031-87883-1_6

Mining generates a very large amount of revenue through taxes, royalties, and dividends, which can be invested in various sectors. Mining companies can take an inclusive approach by working with communities to identify the mining potential through positive and negative impacts. Companies can support a range of local participatory decision-making processes for mining operations, equitable benefit sharing, and dispute resolution (Panossian 2019). Mining can also help build peaceful societies and strengthen the rule of law by preventing and resolving conflicts between companies and communities, respecting human and indigenous people's rights, refraining from corruption, ensuring transparency in reporting revenue streams, and supporting representative decision-making by citizens and communities in extractive industry development (Panossian 2019).

Is the current situation in Mexico's mining sector in terms of its legislation a path to have sustainable development, and taking care of the environment? This is the question that this chapter seeks to answer, based on the Mexican mining industry, its economic development, policies, and in terms of possible carbon sequestration.

2 Mining in Mexico

As mentioned before, mining is the beginning of all productive chains and is present in all economic activities. It is a fundamental industry for Mexico's development, as it provides raw materials for a wide range of industrial activities, generates employment in the areas that are difficult to access, attracts significant investment, also as one of the main foreign exchanges, and contributes to local and regional development (Gobierno de México 2015).

2.1 Importance of Mining

As stated earlier, one of mankind's oldest activities is mining. The relationship between human evolution and development of the mining sector has been reflected in our history. Mineral products are a measure of humanity's progress, and having control over them implies economic development at a global level (Ali et al. 2017; Meinert et al. 2016). Today, it is difficult to point out any activity directly or indirectly not linked to the use of products derived from mining (Johansson et al. 2013). From mobile phones, windows, cars, and daily-need items, mining, to a greater or lesser extent, is present everywhere, as most aspects of our modern life are based on elements, such as metals, industrial minerals, precious metals, aggregates, and coal, among others, to be used in even more complex areas such as power generation, building construction and even most sophisticated technologies used in explorations outside our planet.

Just as mining underpins our modern lives, it underpins much of our economy. It not only provides jobs but also generates foreign exchange and represents an important part of the producing country's GDP (Zheng et al. 2018). Globally, as of 2020, the top 40 companies in the mining industry alone recorded a total revenue of approximately US$655 billion (Garside 2022a), as can be seen in Fig. 1.

Among these leading companies, in 2021, the mining company Glencore ranked first in the list by obtaining approximately US$142 billion in revenues, mainly derived from copper production (1.3 million tonne), as well as zinc and aluminum, among others. BHP is a mining company that ranks second, thanks to its coal production, which reaches around US$60.8 billion (Garside 2022b). Despite these high numbers, the net profit margin, such as the measure to

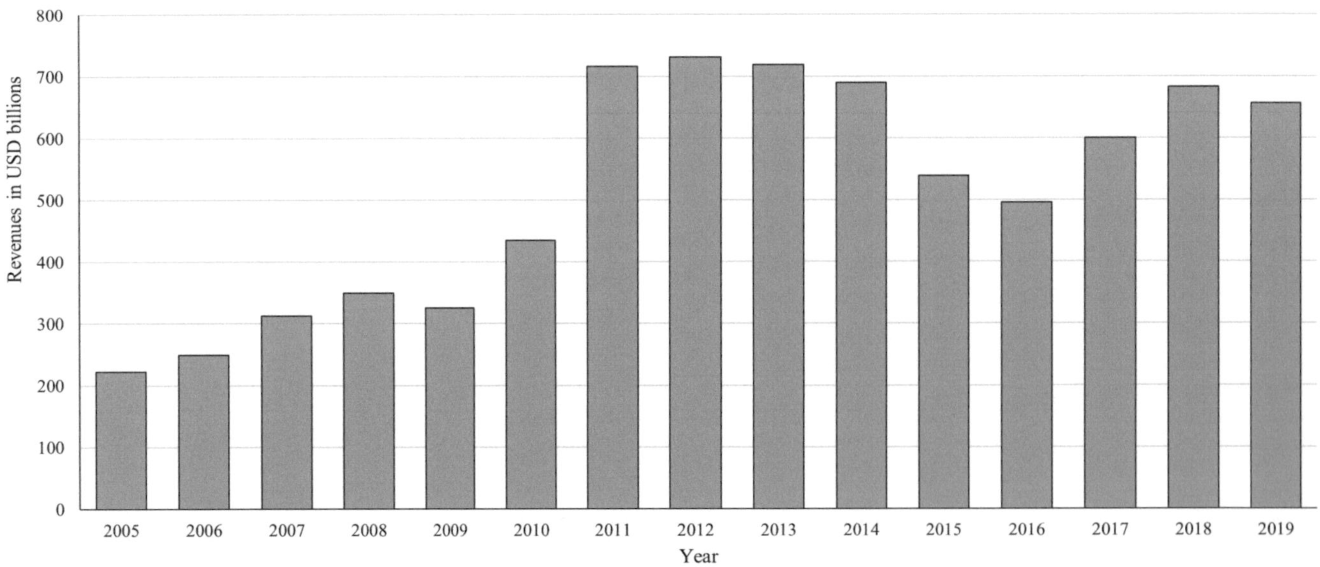

Fig. 1 Revenues of leading mining companies worldwide

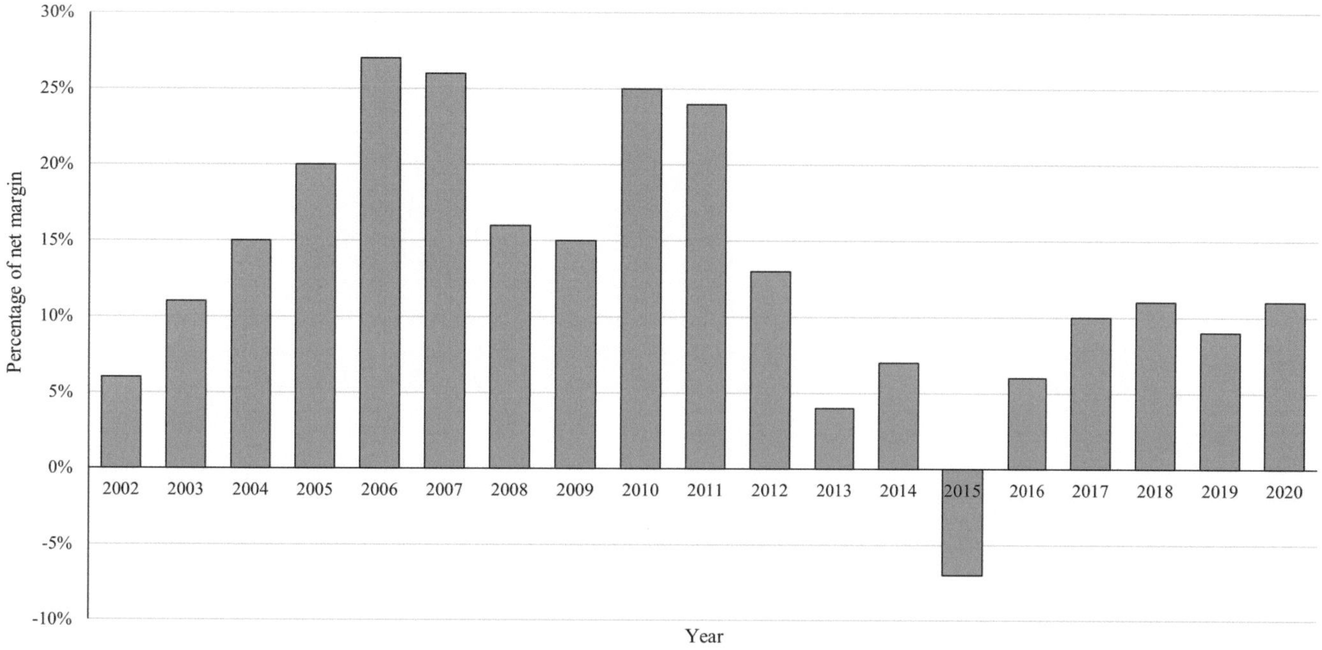

Fig. 2 Net profit margin of major mining companies worldwide

describe a company's profitability, decreased from 24% (2011) to 11% by 2020 (Statista 2021). Figure 2 shows the net profit margin of the world's leading mining companies from 2002 to 2020.

These leading companies earned 11 cents for every dollar they earned (Statista 2021). Undoubtedly, mining is one of the most important economic activities in Mexico; it is the first producer of silver in the world, generating around 187 million ounces in 2019. Mexico is also found to be among the top ten producers of the minerals mentioned in Table 1 (Cámara Minera de México [CAMIMEX] 2020).

At a national level, the mining sector contributes 2.5% of the country's GDP and 8.1% of the industrial GDP (Secretaría de Economía 2022). In addition to its importance as a producer at a global level, its investments are relevant in the areas of social development and environmental protection. In 2019, the investment in these areas was Mexican peso (MXN) 5,327 million (CAMIMEX 2020).

Modification based on data from CAMIMEX (2020).

2.2 Mining Centres

Mexico is one of the main countries in Latin America, and worldwide, to invest in the mining sector mainly because of its diverse geology (Servicio Geológico Mexicano [SGM] 2017). A mineral deposit refers to accumulations or concentrations of a suitable mineral substance to be exploited economically. These deposits are not formed sporadically but developed throughout the planet's geological history in different environments (Misra 2000). There are several types of mineral deposits, such as metallic, precious metals, base metals, ferrous metals, minor metals, carbon deposits, precious and semi-precious stones, or even mineral rocks (Stevens 2010).

In Mexico, it can differentiate certain regions, (as seen in Fig. 3) called metallogenic provinces, which represent deposits of a characteristic type or group of deposits that have attributes suggesting some genetic relationship, such as composition, mineralisation styles, or morphology. It is important to mention that these regions are not defined by a specific size (Misra 2000). Within this context, the Mexican Geological Service has defined the following 12 metallogenic provinces:

1. Silver (Ag)
2. Ag, lead (Pb) and zinc (Zn)
3. Ag, Pb, Zn, copper (Cu), manganese (Mn) and molybdenum (Mo)
4. Ag, Pb, Zn and Mn
5. Stone aggregates
6. Gold (Au) and Ag
7. Au, Cu and Mo
8. Celestite and fluorite coal
9. Chromium (Cr)
10. Iron (Fe) and Au
11. Phosphorite
12. Salt, sulphur (S) and stone aggregates.

Table 1 Main minerals produced in Mexico and their importance worldwide

Overall ranking	Mineral	Global production (tonne) 2019–2020	National production (tonne) 2019–2020	National producer states
2nd	Fluorite	7 million	1.23 million	San Luis Potosi (SLP), Coahuila
3rd	Celestite	220,000	39,215	Coahuila
3rd	Sodium sulphate	15.7 million	745,892	Coahuila
3rd	Wollastonite	700,000	159,498	Sound
5th	Bismuth	19,000	251	Coahuila
5th	Molybdenum	290,000	18,562	Sound
5th	Lead	4.7 million	260, 390	Zacatecas, Chihuahua, Durango
6th	Cadmium	–	850	Coahuila, SLP
6th	Magnesium sulphate	3.2 million	62,222	Coahuila
6th	Zinc	12.9 million	688,461	Zacatecas, Durango, Chihuahua and Edo, from Mexico
6th	Diatomite	3 million	96,400	Jalisco, Muchoacan
6th	Barite	9.5 million	378,295	New Lion
7th	gypsum	140 million	5.70 million	Baja California Sur (BCS), Morelos, Nuevo Leon, SLP
8th	Salt	270 million	8.6 million	BCS, Veracruz, Guerrero
8th	Gold	106 million ounces	3.5 million ounces	Sonora, Guerrero, Chihuahua
9th	Copper	20 million	768,542	Sonora, Zacatecas, SLP, Chihuahua and BCS

Fig. 3 Concentration of mineral deposits in Mexico

Fig. 4 Mining centres in Mexico

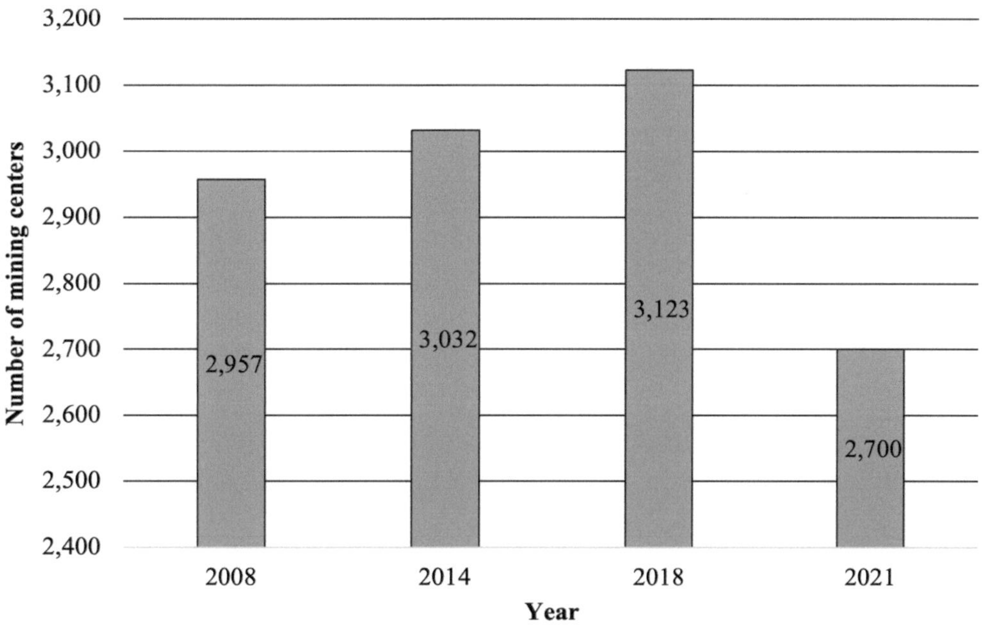

Although the mining sector is present in all 32 states of the country, it is in the north where most of the mineral deposits are concentrated, specifically in Sonora, Zacatecas, Durango, Coahuila and Chihuahua states. Today, there are around 2,700 mining centres (SGM 2021), of which 690 are marked in the states (Secretaría de Economía 2022). However, according to Fig. 4, this number has changed over the years, compared to mining centres in 2018, and the same has decreased.

Figures 5, 6, 7, 8, and 9 present the mining centre's production in Chihuahua, Aguascalientes, Baja California, Baja California Sur, and Campeche. Of the 50 different products that are obtained in the Chihuahua mining centres, only the first 20 with the highest production are shown in Fig. 5, their main product is gold and silver.

2.3 Negative Consequences

In the world, mining involves benefits and risks for the communities living at or nearby the places of these activities taking place. Currently, in Mexico, mining is not only one of the main reasons for social and environmental emergencies but it also has several privileges. At the end of 2018, different reform initiatives regarding the Mining Law were released due to the urgency of a change in the legal framework for its sustainable development, as it was susceptible to damage economical, political, and social areas (Centro Mexicano de Derecho Ambiental [CEMDA] 2019).

Around 70% of the national territory has mining potential, of which 35% was granted concessions between 2000 and 2010; by May 2010, 26,559 mining concessions were provided, which covered approximately 13.8% of the nation's surface (Cárdenas 2013). This number decreased by March 2019, and the mining concessions covered more than 22 million ha of surface area (CEMDA 2019). Some of them have caused social and environmental problems because of concessions granted in protected areas and indigenous regions and ejidos.

Even though mining activity is not allowed in Natural Protected Areas (NPAs), according to the Article 6 of the Mining Law, it is stipulated that exploration, exploitation, and use of mineral resources are of public utility and enjoy priority over any land use or exploration (Bastidas-Orrego et al. 2018). Under this context, mining has access not only to territories that are part of ejidos and communities but also to NPAs (Hernández and Navarro 2020).

This condition has triggered several indiscriminate concession permits. Some of them have been identified and documented within the National Inventory of Contaminated Sites of the Ministry of Environment and Natural Resources (SEMARNAT, by its acronym in Spanish), which not only points out mining as the main cause of severely contaminated sites but also registered 62 mining projects located in NPAs. Likewise, the Contaminated Sites Information System (SISCO, by its acronym in Spanish) noted 632 potentially contaminated areas, of which 84 belonged to mining, while 11 of them are part of NPAs (Gómez Godoy and Peláez Padilla 2020).

There were 1,609 mining concessions covering an area of 1,486,433 ha up to 2010 whereas in most of the NPA categories (except for natural monuments), mining areas coincided with concessions. With 75% within the Natural

Fig. 5 Mining centre in Chihuahua

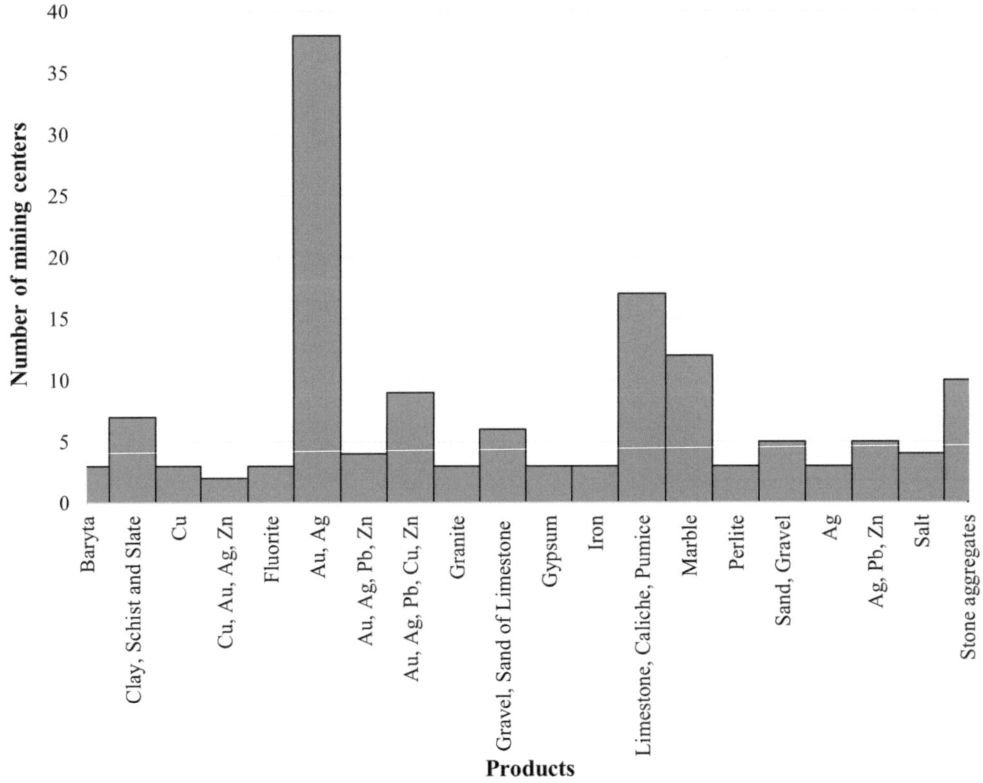

Fig. 6 Mining centres in Aguascalientes

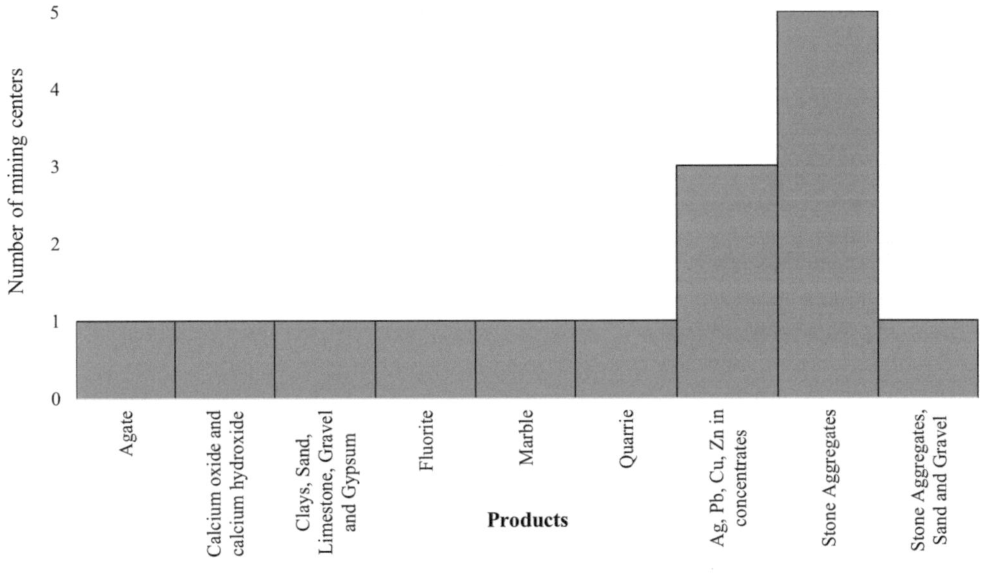

Resources Protection Area (NRPA), 47% in the Flora and Fauna Protected Area (APFF), 63% in Biosphere Reserve (BR), 15% in National Park (NP) and, finally, 22% in Sanctuaries (S) (Armendáriz-Villegas et al. 2015).

Damage generated by the mining industry is not only observed at the beginning or end of its operations but also throughout the different stages of the industry. In the development, production, and closure stages, more sophisticated equipment is used for soil drilling, smelting, and refining of products, and consequently, in these phases, the effects on the environment and the population are more perceptible.

In the extraction phase, open-pit mining exposes minerals to erosive and meteoric agents without being leached (Martinez 2003), that is, without the chemical process in which the mineral of interest is extracted from other substances that do not have an economic value (AZO Mining 2014). Leaching is a technique that unleashes great risks, for example, heap leaching consists of spraying thousands of tonnes of extracted material with a sodium cyanide solution

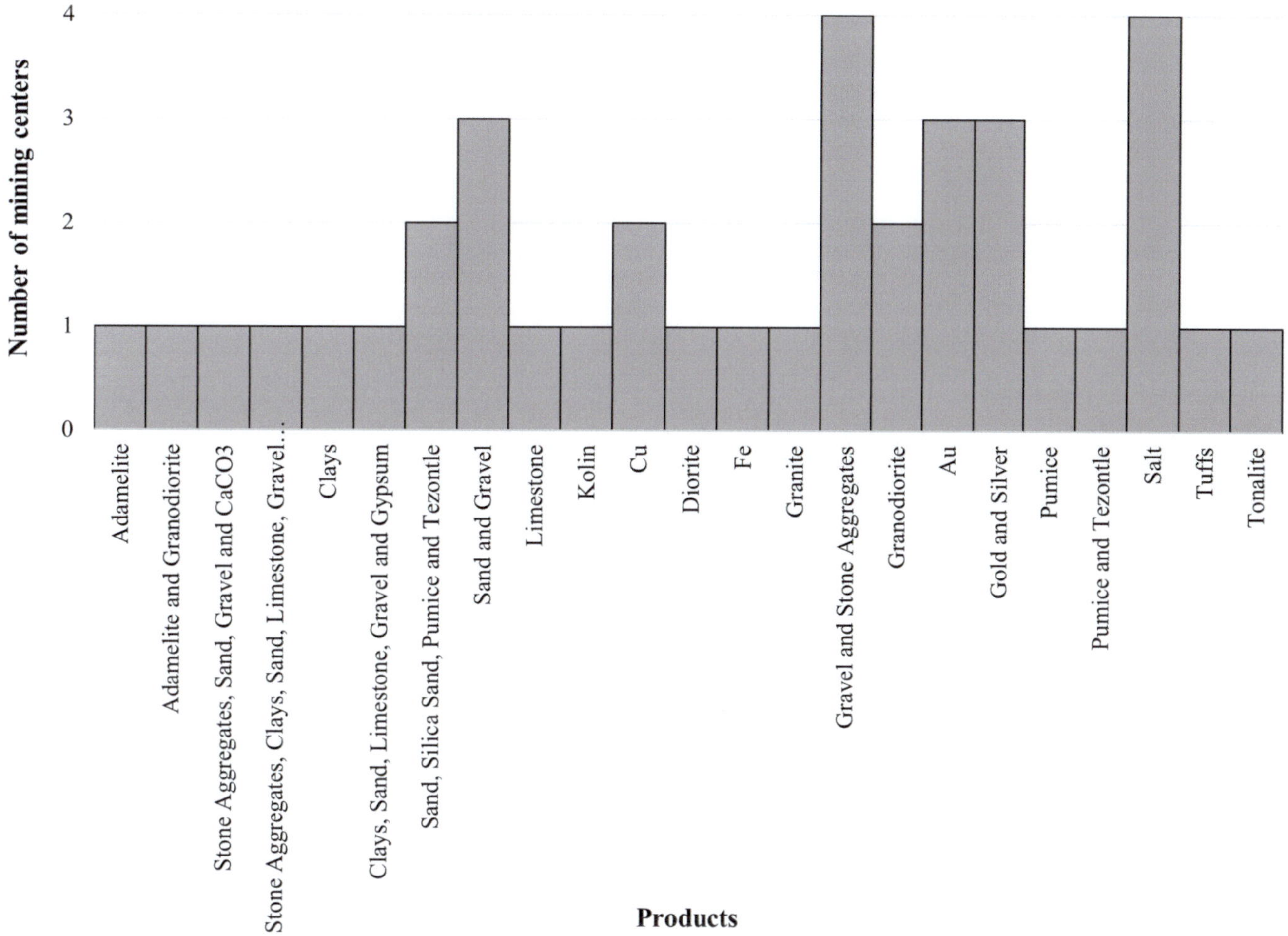

Fig. 7 Mining centres in Baja California

Fig. 8 Mining centres in Baja California Sur

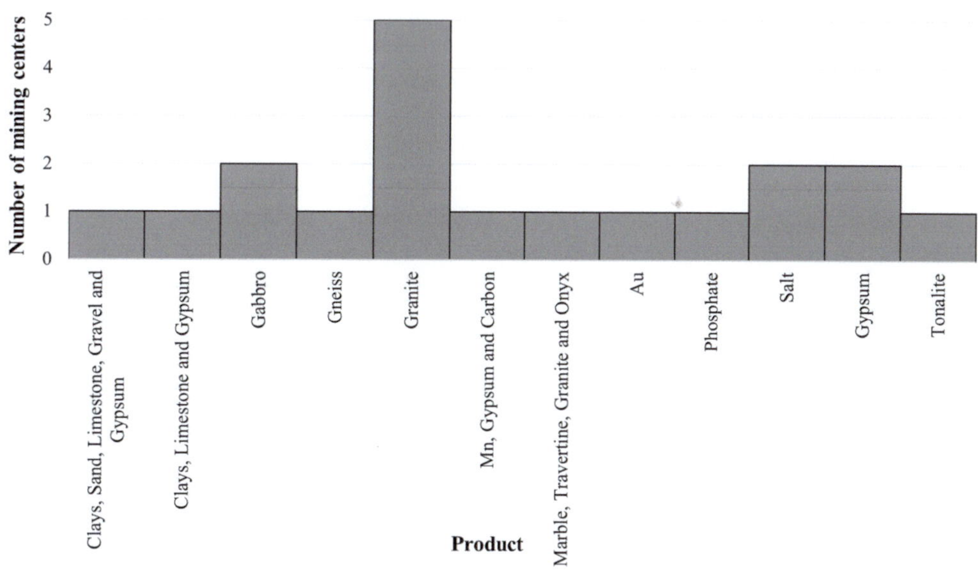

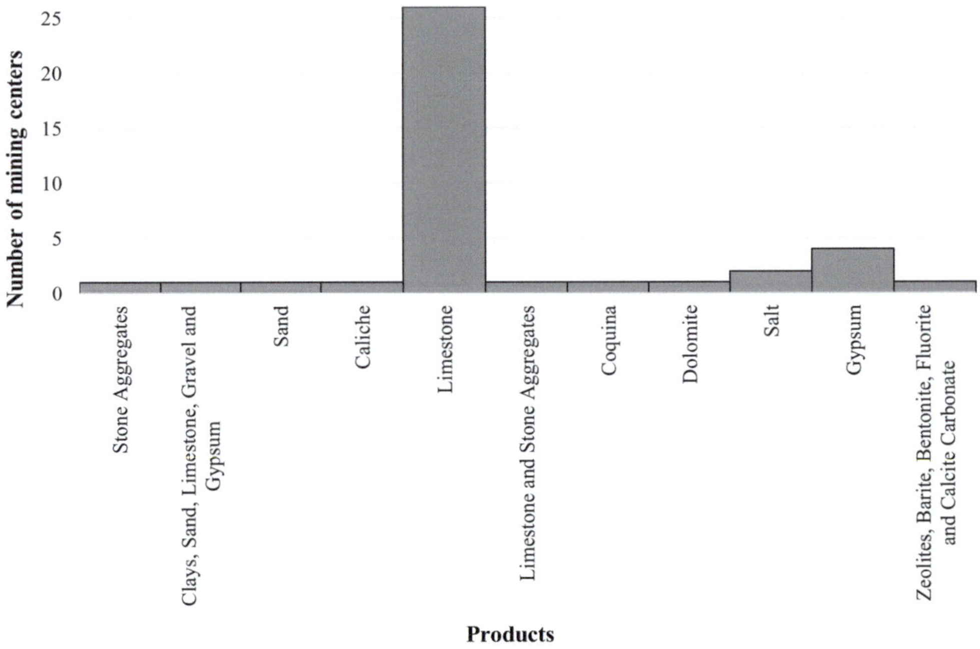

Fig. 9 Mining centres in Campeche

to obtain gold. This technique, mostly used in open-pit mines, involves high ecological impacts, recorded as environmental disasters (Cárdenas 2013).

On the contrary, in the final phase, when metallurgy or refining processes are applied to the minerals, they result in air, soil, and water pollution (Martínez 2003). Mining tailings are wastes derived from the final processes applied to the minerals, and poor management of these can lead to acidic water sources due to the chemicals used in the process (Martínez-Castilla 2003).

Likewise, the amount of water used in mining is alarming. In 2014, 437 million m^3 water was extracted for this activity; this amount could supply the human needs of at least four states in the country. Approximately 50% of this water comes from three states: Zacatecas (55.8 million m^3/year), Michoacán (44.4 million m^3/year), and Sonora (107.9 million m^3/year) (Llano 2016). The excessive use of water for leaching affects the quality and its availability, 38% of mining projects are in aquifers classified as overexploited (Ramírez 2017).

The case of the Herrero de la Bufa mining company in Durango is one of the many examples where dams for tailings have caused water pollution (Ramírez 2017); however, the worst environmental disaster in recent years in Mexico occurred in 2014, in Sonora, where 400,000 m^3 of copper sulphate acid was spilled, which expelled concentrations of heavy metals, such as aluminum and zinc, on waterways near Cananea. This disaster affected around 24,000 people and amounted to US$134 million to repair. The Federal Attorney's Office for Environmental Protection (PROFEPA, by its acronym in Spanish) reported that this accident was the result of open pipes, and the water from the mine tailings was expelled into the Bacanuchi and Sonora Rivers (Wilton 2015).

In Mexico, mining is the industry that has claimed most of lives. According to Castro-Soto (2013), in his excerpt from the chapters of the popular manual *La Mina nos Extermina*, in 2012, 25 workers died in the Coahuila coal region. In 2008, a spill of mining tailings from the La María mine was reported, affecting flora, fauna, and the rainwater network. In 2010, the MINDFANDERS mining company, which illegally operates the Dolores mine in Chihuahua, caused an uncontrolled sodium cyanide spill. In 2012, carbon monoxide (CO) poisoning was reported at the El Mono mercury mine in Querétaro, resulting in three deaths.

In case of regions concerning indigenous people, the legislations are present. However, legislations have certain limitations; for example, the General Law of Ecological Balance and Environmental Protection (LGEEPA, by its acronym in Spanish) allows regions to participate in the Environmental Impact Assessment. However, one of the exceptions is mining activity; this fact makes visible the inequality when it comes to passing mining regulation in their territories (Hernández et al. 2018). In Mexico, there are 494 indigenous municipalities, of which the state of Oaxaca alone has 245, within these, in 2016, 319 mining concessions were provided (Hernández et al. 2018).

In 2019, approximately 7,100 indigenous community displacements were registered, including those derived from the contention with mining companies in their quest to extract mineral resources. The Mexican Commission for the Defense and Promotion of Human Rights (CMDPDH, by its acronym in Spanish) notes that in 2017, around 20,000 people were affected, triggering their displacement within Mexico, of which 60% were indigenous people (Murray 2020).

The negative effects on community health because of mining activities are diverse; gestational diabetes mellitus develops from exposure to metals and metalloids, such as magnesium, arsenic, and mercury, among others, through water, air, and even food (Cortez-Lugo et al. 2021); pneumoconiosis (lung disease) of different types is caused from exposure to coal, silica, and iron oxide, among others; and respiratory, ophthalmological and nervous problems are caused by noise, dust, and changes in the landscape, among other diseases (Martínez-Castilla 2003); however, mining has also claimed lives around the world. For example, in Zacatecas, between 2009 and 2013, approximately 100 people were killed due to poor safety conditions and health problems caused by the mining industry. In Coahuila, in 2012, 25 workers died; all the same, between 2006 and 2011, coal mining was responsible for 147 deaths because of accidents in pits and subway mines (Castro-Soto 2013).

There is a preference in mining allocations over the rights of peoples and communities, in addition to the fact that mining companies have identified strategies on how to access, such as corruption, through bribing of authorities, dividing the community, and repression.

Despite its lucrativeness, mining does not always translate into benefits for the communities where it operates. On occasions, mining companies have acquired land in ejidos and agrarian communities for relatively low amounts, despite their high income. This has led to the fact that, despite the business profits, this is not reflected in these communities. Another example is the communities where gold and silver are extracted, which, far from being places that are benefitted from mining activity, are in dire conditions, such as poverty and insecurity. In 2015, although gold production was abundant in the four municipalities of Zacatecas, Sonora, Chihuahua, and Guerrero, people of these regions suffered from extreme poverty (de la Fuente López and Olivera Villa 2017).

2.4 Activities Carried Out to Remedy Pollution and Prevent Damage to the Environment

The environmental impact generated by mining activity results in dimensionless damages if it is not formulated under a regulatory framework that works correctly (Secretaría de Economía 2022).

In general, the environmental impact resulting from mining is related, depending on each phase that makes up a mining project, which is standardised by LGEEPA together with the Law of National Waters (LAN, by its acronym in Spanish), complemented by the Mining Law (Hernández et al. 2018).

To begin the initial prospecting stage and conduct further exploration, obtaining authorisation from the federal executive through a mining concession, which is valid for 50 years, is essential (Centro de Estudios de las FinanzasPúblicas (CEFP) 2018). At this stage, an area of interest is identified, and a detailed description is made of its geological characteristics, background, and all the legends that suggest that this area could become a mineral deposit. The presence of SEMARNAT is not necessary at this stage, because the studies developed on it do not cause an environmental effect (Secretaría de Economía 2022).

The exploration stage consists of evaluating mineral resources for exploitation, in addition to having the mining concession; at this stage, it is necessary to make a preventive report or an Environmental Impact Statement (MIA, by its acronym in Spanish) based on area studies and the degree of impact that would trigger such an activity. The MIA documents the actions to avoid or reduce negative impacts. Conversely, if the area in question is located within an NPA, authorisation must be obtained from the National Commission for Protected Areas (CONANP, by its acronym in Spanish) (Secretaría de Economía 2022).

Stage three consists of the mining project development, which includes infrastructure plan, mineral reserve studies, and production costs. At this point, it is essential to have SEMARNAT's environmental impact consent, for which the MIA is submitted and, if necessary, an Environmental Risk Assessment (ERA, by its acronym in Spanish) is included, provided that the mining activities are considered highly hazardous (Environmental Law Institute 2000). For the latter to be approved, an accident prevention programme must be submitted, outlining the measures and actions to prevent the disasters mentioned in ERA. Not only do these standards exist to ensure a low-risk mining activity but other standards regulate leaching systems and water use. For example, there are Norma Oficial Mexican (NOM) or standards, such as NOM-001-SEMARNAT-1996, which refers to the maximum permissible contaminant limits in wastewater discharges (Secretaría de Medio Ambiente y Recursos Naturales 2003); NOM-141-SEMARNAT-2003, which refers to the construction and operation procedure related to dams for tailings (Secretaría de Medio Ambiente y Recursos Naturales 2004). All the same, water use must also have a concession granted by the National Water Commission (CONAGUA, by its acronym in Spanish) (Secretaría de Economía 2022).

A mine is in its production stage when the material of interest is already being extracted; it also includes product treatment and refining processes. Finally, when the productive life of the area ends, actions are taken to reduce the environmental and ecological impacts of the mining activity.

Primarily, all the equipment used, such as machinery, and the processing plant, among others, are removed from the site. The final restoration phase also begins, where the regulations regarding deposits and leaching of tailings are enforced.

Finally, the areas used are treated to be repopulated with native species of flora and fauna. As for atmospheric contamination, chemical levels in the air, water, and air are monitored not only at the surface level but also at the subway level. As part of the remediation programme for contaminated sites, measures to eliminate or reduce contaminants are based on NOM-147-SEMARNAT/SSA1-2004. This standard establishes criteria to determine the soils' remediation concentrations contaminated by arsenic, barium, beryllium, cadmium, hexavalent chromium, mercury, nickel, silver, lead, selenium, thallium, and/or vanadium (Secretaría de Economía 2022).

In most cases, the area of interest is restored after the exploration stage because the expected deposit was not found; in this case, activities, such as reforestation, recovery and restoration of soils, and adaptation of flora and fauna species that were initially present in the area, are carried out (Saavedra and Hernández 2015). Today, although environmental management systems are not required by law, they are implemented by certain mining companies. These systems enhance mining operation's efficiency, and, in turn, avoid large amounts of pollution, and the ISO 14000 standards adaptation as norms that cover environmental aspects, provides investors with greater security and attraction. In Mexico, the mining industry can participate in a voluntary management programme, which grants economic incentives or recognition. An example of this is Minera Carbonifera Río Escondido, which in 1997 obtained the ISO 14001 certificate for environmental quality, the first mining company in the country to receive this recognition (Environmental Law Institute 2000).

The Priority Toxic Substances Environmental Management Program and the Program for the Reduction and Integrated Management of Hazardous Industrial Waste (1996–2000) were programmes promoted by the National Ecology Institute to encourage the voluntary adoption of environmental prevention measures (Environmental Law Institute 2000). The evasion of regulations mentioned throughout this document carries criminal, civil, and administrative penalties. In case of damages to third parties derived from mining operations, penalties include imprisonment ranging from 3 months to 6 years, in addition to fines ranging from 1,000 to 20,000 days of minimum wages. Penalties can also lead to a suspension or even demolition of the works that caused damages, among other sanctions (Environmental Law Institute 2000).

Australian Mining Equipment, Technology and Services (METS) companies leverage their strong investment in research and development to increase productivity, improve safety, and reduce the environmental footprint of mining projects, contributing to sustainable mining practices. In this regard, METS supports Mexico in the sustainable mining sector in automation, exploration, engineering, mineral processing, environmental management, community relations, safety, research and development, education, and training areas. This contributes greatly to sustainable mining practices and mine safety. The Mexico-based companies assist to identify the most appropriate technology, equipment, or services, with a focus on sustainability (Parlamaint of Australia 2015).

2.5 Development of Public Policies in Mining Industry

The Mexican mining history boomed in the nineteenth century with the beginning of legal bases for its management. The culmination effect of the Spanish Empire mandated around New Spain, because, under its dominion, the empire's wealth was based on Mexico's and other country's minerals (Herrera and Parra 2018).

Approximately, before 1770, the mining sector was subdued by the state, imposing large taxes and controlled the inputs, such as quicksilver and gunpowder, to maintain production. Later, in 1777, the General Mining Tribunal was created, which enacted a Bank of Provisions and decreed the New Ordinances to give autonomy in 1778; however, it was not until 26 March 1783 that these were ordered to be complied with (Contreras 1996).

The New Ordinances were made up of 19 titles. The first one referred to the court constitution; the second section was about Royal Mines and their judges and deputies; the third, fourth and fifth titles referred to the sector jurisdiction. The sixth and seventh titles described how and by whom the mines could be acquired; the eighth title described dimensions of the mining areas, and the ninth title reported how the mines had to be 'worked, fortified and protected'. The tenth section referred to drainage mines and other titles, among the most important, dealt with consumption, the Bank of Mine Additions, the College of Mines, and mining privileges (Contreras 1996).

Despite the independence war ending panorama, there were new mining policies emerging, and the mining development programmes continued to be applied to lower taxes and respect mining rights (Herrera and Parra 2018), in addition to stimulating foreign investments.

In 1822, the Provisional Board decreed an only pay 3% on the value of metals, whether gold or silver, and only a year later the collection about the same was regularised in addition to noticing the considerations in which foreigners could acquire mines in the country (Martínez-Castilla 2003).

In 1826, the Mining Establishment was established, which annulled the General Mining Tribunal, but years later, in 1842, it was replaced by the Mining Development and Administrative Board to protect the so-called quicksilver mines and to adjust imposed duties (Herrera and Parra 2018).

In April 1853, the Ministry of Development, Colonization, and Industry was established to promote not only mining activity but also agriculture and manufacturing industries, in addition to new transportation and communication routes (Herrera and Parra 2018).

The Mining Fund expired in January 1861, giving rise to the Public Credit (Martínez-Castilla 2003). Thirty years later in February 1883, the Mexican Mining Society was established; among its proposals was the need to publish mining charts derived from the main mining districts of the country to concentrate information in order to encourage foreign or national investment (Martínez-Castilla 2003).

The titles of the New Ordinances remained until 1884, when the Mining Code was established after Durango and Hidalgo states sought to update said titles through Special Mining Codes. These states through Codes decreed that the mines corresponded to the state and that the product obtained was only the owner area property. With this, article 27 was taking shape (Martínez-Castilla 2003), and offered facilities for foreign investment.

After creation of the Special Mining Codes, a mining law was enacted in June 1892 that proposed consolidation in mining property terms. Under it, the miners obtained the properties they wanted through a title granted by the Secretary of Development after paying a federal tax, through a mining concession. However, the area owners were free to exploit without a concession. After that, the time for which an area was granted varied; however, this mining law and its regulations established principles that differed from those of the New Ordinance (Martínez-Castilla 2003). After the mining law, Porfirio Díaz in 1909 decreed a law in which it was mentioned that minerals were the nation's domain, but conversely ruled that they were the exclusive property of the area owner (Cárdenas 2013).

It was not until 1917 that, together with the Constitution, a period of regulation began, where Article 27 was decreed, which described the nation's direct dominion over mineral resources. This year, the Ministry of Development was also again organised to establish the Ministry of Industry, Commerce, and Labor (Auditoría Superior de la Federación [ASF] 2015).

Nine years later, in 1926, the Law of Mineral Industries was approved. This law established that only Mexicans would be granted mining concessions and could take advantage of the country's resources, and for no reason it was possible for foreign companies to obtain mining concessions. Moreover, any industry in this sector must be composed of at least 90% of national workers and the concessions would cover a maximum of 30 years; in addition, these were granted in stages and were of four types: exploration, exploitation, transportation, and storage. This law was modified in 1930 in favour of national reserves, and in 1935, the Regulation for the Exploitation of Mining Reserves came into force. In the same year, the Secretariat of National Economy was created to control extractive industries (ASF 2015).

The Mining Law had more than one modification; in 1961, its variation required that participation of Mexican capital in the companies must be at least 51%. There was also a change regarding concessions because they could cover 25 years with indefinite extension; however, if it was proved that work exploitation was involved, then the company was no longer granted concessions (ASF 2015).

In 1974, the Trust 'Non-Metallic Minerals Mexicanos' was created to improve the industry, but in 1990, it was modified with the name of Mining Development Trust Fund (FIFOMI, by its acronym in Spanish) to not only focus on non-metallic but also on all varieties of elements, except for radioactive minerals and petroleum derivatives. The Secretariat of Energy, Mines and Parastatal Industries (SEMIP, by its acronym in Spanish) was in charge of FIFOMI (ASF 2015).

More laws were enacted in the mining sector; for example, in 1988, the LGEEPA was made public to protect the ecological balance of the country. As a result of the negotiations for the North American Free Trade Agreement (NAFTA), in 1992, the current Mining Law was promulgated, which regulated all mining activities in the country. This law determined mining concessions and decreed that all minerals located in the Mexican territory could be exploited by individuals through a concession (Mining Mexico 2013). Also in 1992, PROFEPA was created to promote sustainable development.

In 1995, the Ministry of Commerce and Industrial Development (SECOFI, by its acronym in Spanish), now the Ministry of Economy, was created to replace SEMIP (ASF 2015). This was exercised to oversee the enforcement of mining policy laws, generate new strategies for new investments, and maintain the existing ones.

More recently, in 2001, SEMARNAT opted for a design, through which it administered mining projects and their possible consequences on the environment through the Environmental Impact Assessment (EIA). It comprised the General Directorate of Environmental Impact and Risk (DGIRA, by its acronym in Spanish) and the General Directorate of Energy and Extractive Activities (DGEAE, by its acronym in Spanish) (ASF 2015).

In 2021, the Mining Law was again modified. This time, the administrative part was simplified with less waiting period for a response and extended the period of a mining concession. If concession was granted, the Mining Law allowed its use for 50 years, with a possibility of another 50 years (Article 15) (Amazar 2018). In 2014, this law was changed for the last time, and included actions regarding environmental protection (ASF 2015).

In 2013, the Federal Law of Rights was changed to apply the payment of 'Special Right on Mining', where the owners having concessions must pay annually an additional tax to companies that sell gold and platinum. This intends to create a 'Fund of Contributions for the Sustainable Regional Development of the Mining States and Municipalities' (ASF 2015).

Since 1992, SEMARNAT, PROFEPA, and the Federal Law of Metrology and Standardisation have issued the following eight Mexican official standards in different years concerning the mining activity:

1. NOM-120-SEMARNAT-2011: Indicates specifications for environmental protection in the areas having dry climates or where vegetation, such as tropical forest and grassland, among others, grows up.
2. NOM-141-SEMARNAT-2003: Refers to management of tailings.
3. NOM-147-SEMARNAT/SSA1-2004: Mentions criteria for the characterisation and concentration determination for soil remediation contaminated by certain minerals.
4. NOM-155-SEMARNAT-2007: Shows the management of leaching systems.
5. NOM-157-SEMARNAT-2009: Contains indications for mining waste management.
6. NOM-159-SEMARNAT-2011: Indicates environmental protection requirements for copper leaching systems.
7. NOM-023-STPS-2012: Indicates safety requirements for risk prevention among workers in both subway and open-pit mines.
8. NOM-032-STPS-2008: Indicates safety conditions and requirements for the installation and operation of subway coal mines.

However, the activities in hydrocarbon production have similarities, but it is not the same situation in policy terms. Hydrocarbons exploration was exclusively under the charge of State through Petroleum Mexicanos (PEMEX) from 1938 until 2013. In 2015, with the Energy Reform, the private sector participation was allowed through contracts for both exploration and exploitation, which triggered a series of changes to facilitate private participation. In 2016, PEMEX maintained its position as the leading producer of hydrocarbons, while mining, mostly in the hands of the foreign private sector, was limited by less consistent regulation. This disparity is largely due to differences in regularity and ownership structures between both sectors (de la Fuente López and Olivera Villa 2017).

Also in 2013, the Mining Fund for Sustainable Regional Development was created; by 2014, the Fund for Hydrocarbon Producing Entities and Municipalities (FEMPH, by its acronym in Spanish) was created, and in August 2014, the Hydrocarbons Law was approved, which could broadly force landowners to lease their properties to petroleum companies; however, this was a contradictory Law in itself, because it gave the Ministry of Energy (SENER, by its acronym in Spanish) the task to guarantee the affected communities rights to use prior, free and informed consultations (de la Fuente López and Olivera Villa 2017). In 2015, the Federal Law for the Prevention and Punishment of Crimes Incurred in Hydrocarbon Matters was approved (de la Fuente López and Olivera Villa 2017). Figure 10 shows the development described above.

2.5.1 Current Legislation

Today, mining activity in Mexico is governed by Article 27 of the Constitution, from which the current Mining Law was created on 26 June 1992. This law determines mining concessions and edicts that all minerals located in the Mexican territory are nation's property and can be exploited by individuals through a concession (Mining Mexico 2013). The Mining Law is complemented by the Agrarian Law, the Federal Law of Rights, the Fiscal Code of the Federation, the Fiscal Coordination Law, the General Law of Ecological Equilibrium and Natural Resources, the Law of National Waters, and the Federal Labor Law (CEFP 2018). Unlike other countries, the Mexican Government does not have any agency to carry out exploitation activities of mineral resources; however, in terms of exploration, the Mexican Geological Service is in charge of these activities (Castillo 2020).

A mining concession is an administrative act through which public administration grants rights to individuals to explore and/or exploit an area of the State. This is an act that takes relevance because it is of outstanding interest from economic criterion. Given the economic difficulty of the State to carry out the projects of this nature, it empowers individuals to do the same, which allows them to have income through taxes generated from the mining sector (Castillo 2020).

Even though the mining sector makes a great contribution to the country's economy, it has lost importance at the international level because of the legal uncertainty it is currently undergoing. Since the beginning of the 6-year term of the present president of Mexico, mining concessions have not been granted and, as a consequence, foreign investments have been reduced. Following example states drop in investment around mining exploration: in 2014, foreign investments were $856 million, while in 2019, these came down to only $435 million; not only that, in 2020, among 77 nations, Mexico was placed at No. 42 in the global ranking of countries to invest in the mining sector (Suárez 2021).

Fig. 10 History of policy development for the mining industry in Mexico

Before the concessions ceased, most of the states in the Mexican territory already had some mining permits. For example, in December 2020, Aguascalientes had a total of 213 mining concessions. In 2019, Campeche had only 17, Chiapas 82, Coahuila 1,967, Durango 3,496, Guerrero 631,405, and Chihuahua had 3,348 mining titles. Chihuahua has national mining exploration assignments protected by the Mexican Geological Service; once this is concluded, disincorporation would transfer mining rights to the private sector (SGM 2020).

Although all Mexican states are currently under the same mining legislation, the situation is not the same for diverse regions that can be extracted in the country. In recent months, Sonora has been highlighted with the projects of presumed lithium deposits, an element having exponential demand in recent years for its essential use in batteries. Based on the existing lithium deposits in Sonora, President López Obrador proposed an initiative to nationalise this resource through the creation of a new public body. This organisation would be in charge of lithium exploration without resorting to concessions, permits, contracts, or authorizations, in addition to establishing reserve zones for its protection. (Chávez and Usla 2022). On 19 April 2022, the Mining Law reform was approved by the Senate (Carrillo 2022); however, the legal loopholes and certain historical precedents have created uncertainty regarding this new reform (Ávila 2022).

2.5.2 Regulation in Terms of Generated Pollution

Contrary to popular belief, the regulations governing the country's mining sector are strict and comply with international standards. The regulations of Mining Law comprise six official Mexican standards (as mentioned in Sect. 3.1) exclusively for environmental protection. These standards deal

Table 2 Ozone air limit values according to NOM-020-SSA1-2020

Limit value	Year 1		Year 3		Year 5	
1 h	0.090 ppm	176 $\mu g/m^3$	0.090 ppm	0.090 ppm	0.090 ppm	176 $\mu g/m^3$
8 h	0.065 ppm	127 $\mu g/m^3$	0.060 ppm	118 $\mu g/m^3$	0.051 ppm	100 $\mu g/m^3$

Table 3 Carbon monoxide (CO) air limit values according to NOM-021-SSA1-2020

Limit value	CO (ppm)	CO ($\mu g/m^3$)
1 h	26.0	30,000 $\mu g/m^3$
8 h	9.0	10,000 $\mu g/m^3$

with direct mining exploration, deposit of tailings, soil remediation, gold and silver ore leaching, mine waste management plans, and copper leaching, in addition to standard NMX-AA-132-SCFI-2006 for soil sampling.

These standards are applied throughout all mining projects, with other laws and by all agencies. At an initial stage, during searching and exploration, an activities report is required, which must comply with standard NOM-120-SEMARNAT-2011. This NOM establishes protection indications for mining activities (R. del Bosque 2021). Prior to extraction and even exploration, an MIA is prepared and submitted under LGEEPA that aims to prevent, mitigate, and restore damage to the environment caused by mining activities (Procuraduría Federal de Protección al Ambiente 2019).

Conversely, for the final stage of operation, standards NOM-155-SEMARNAT-2007 and NOM-159-SEMARNAT-2011 are applied for gold, silver, and copper leaching activities, respectively. Here, the waste is identified by its degree of danger, in addition to specifying the environmental protection requirements, during characterisation and site preparation stages at the location of leaching (Procuraduría Federal de Protección al Ambiente 2016).

At this stage, it is necessary to develop a hazardous waste management plan under NOM-157-SEMARNAT-2009 and NOM-141-SEMARNAT-2003 guidelines, in addition to an accident prevention programme (R. del Bosque 2021). These NOMs comprise instructions to characterise mining tailings and the criteria for site preparation (Secretaría de Gobernación [SEGOB] 2004). For soil remediation resulting from mining activities, NOM-147-SEMARNAT/SSA1-2004 is affected to establish criteria for soils contaminated by arsenic, cadmium, barium, mercury, and chromium, among others (SEGOB 2007).

Regarding air emissions, mining activities are subject to national standards. A unique environmental license, as well as a Pollutant Release and Transfer Report, is required. However, recent drafts of Mexican official standards point to four other NOMs, as shown in Tables 2, 3, 4 and 5.

Table 4 NO_2 air limit values according to NOM-021-SSA1-2020

Limit value	NO_2 (ppm)	NO_2 ($\mu g/m^3$)
1 h	0.106	200
8 h	0.021	40

Table 5 Limit values for lead in air according to NOM-026-SSA1-2020

Limit value	NO_2 ($\mu g/m^3$)
Annual	0.50

3 Public Policies for Carbon Sequestration

Globally, the largest source of carbon dioxide (CO_2) is emissions from fossil fuel (coal, oil, and gas) combustion; these fuels also produce other gases, such as methane and carbon monoxide. All these gases are known as greenhouse gases (GHG), which allow solar radiation to pass through but do not allow heat to escape and remain locked in. Although GHGs are necessary for the survival of human life on the planet, in excess they cause what is known as global warming (Singh 2013). According to Mexico's National Emissions Inventory published in 2015, GHG in 2013 reached 665 metric tonne (Mt) CO_2 equivalent, with 75% being CO_2 emissions or 449 Mt (World Bank 2016).

The country's government has been trying to cope with the consequences of climate change through measures and regulations, such as granting renewable energy certificates to projects producing electricity from renewable sources as well as those who use clean energy; however, these efforts worldwide have not been enough to cope with mounting GHG pollution (Domínguez and Scheffler 2022).

One way to mitigate climate change effects was the introduction of the Energy Law Reform in 2013. It required producing 35% of Mexico's electricity from clean sources by 2024. However, by 2020, only 25% of electricity was produced from clean sources (Domínguez and Scheffler 2022).

Before the 2013 reform, activities of the energy sector were state-owned and carried out by state-affiliated entities, such as PEMEX and Federal Electricity Commission (CFE, by its acronym in Spanish). However, after 2013, these became the responsibilities of SEMARNAT and PROFEPA. Later, in 2014, initiatives were made to implement the 2013 Constitutional Reform. Deduced from this, the federal executive imposed different entities, such as the Mexican Agency for Safety, Energy and Environment (ASEA, by its acronym in Spanish), in addition to instructing the following new laws enacted in August 2014 (World Bank 2016):

- Law of Coordinated Regulatory Bodies of the Energy Sector
- Electricity Industry Law
- Law on the Federal Electricity Commission
- Geothermal Energy Law
- Law on Hydrocarbons
- Law on Hydrocarbon Revenues
- Law on the Mexican Petroleum Fund for Stabilization and Development (Mexican Petroleum Fund for Stabilization and Development)
- National Industrial and Environmental Safety Agency Law for the Protection of the Hydrocarbons Sector ('ASEA Law'), and
- Petróleos Mexicanos Law.

In addition, the following laws were altered (World Bank 2016):

- Federal Law on Duties
- Federal Budget and Treasury Accountability Law
- Federal Law on Parastatals
- Foreign Investment Law
- General Law on Public Debt
- Law on Acquisitions, Leasing and Services of the Public Sector
- Law on Public Works and Services related to the same
- Mining Law
- National Water Law
- Organic Law of the Federal Public Administration
- Law on Public-Private Partnerships, and
- Fiscal Coordination Law.

Mexico has a General Law on Climate Change, which regulates atmospheric emissions. This is developed through the National Climate Change Strategy and the Special Climate Change Programme. The former was created by SEMARNAT, which sets the national medium and long-term climate change policy and dictates necessary actions to develop climate change-related projects (World Bank 2016).

3.1 Carbon Capture

One way to reduce CO_2 emissions into the atmosphere is carbon capture, utilisation and storage (CCS), which aims to provide a regulatory framework for CO_2 in Mexico (World Bank 2016). CCS consists of three processes: CO_2 capture, transport, and storage (Singh 2013). This technology encompasses different processes to reduce large-scale emissions of carbon in the atmosphere by capturing CO_2 and storing it underground safely and sustainably. Thus, it is considered a 'notional decrease', as it does not reduce the amount of CO_2 produced but only cuts down its impact (Secretaría de Energía 2015).

The carbon-capture process consists of separating CO_2 from some other gaseous substances, because the smoke coming from power plants contains only 10–12% of CO_2. This is an expensive process and is carried out by three methods: post-combustion separation, oxy-fuel separation, and pre-combustion separation (Singh 2013).

The second step involves bringing the stored CO_2 to a particular site by ship, pipeline, rail or even by truck, although pipeline transport is the most viable transport method. Finally, for its storage, a geological formation capable of safely retaining CO_2 in the long term is required. This may include basaltic formations, deep saline aquifers, inaccessible coal seams, and depleted oil and gas reservoirs, among others. (Singh 2013).

The mining sector could play an important role in achieving carbon sequestration through rocks (especially ultramafic rocks comprising magnesium silicate). These are commonly found in tailings, and they can extract CO_2 in a natural manner and encapsulate it, generating new carbonate minerals. In this way, carbon can be permanently stored. For example, during chemical weathering of mine tailings, hydroxide minerals react with CO_2 and generate magnesium carbonate, which is a cement-like product and has the potential to store carbon for thousands of years (Williams 2020).

3.2 Strategies for Carbon Sequestration in Mining

The term 'carbon sequestration' is used to describe processes that seek to remove CO_2 from the atmosphere or drive it into sinks in oceans or terrestrial environments (USGS 2008). Some strategies to achieve decarbonisation in the mining sector are renewable energy installation, such as solar, wind, or battery power, which results in the saving of around 25% of electricity costs; this also implies better conditions, in marketing terms, of mineral products, and low carbon content (Touchpoint Marketing 2020).

Under the Kyoto Protocol agreements within the United Nations framework on climate change, some relatively low-priced activities for carbon sequestration were described, such as afforestation, where non-forest land is converted to forest, reforestation, improved agricultural activities, and could include improved cropland practices with a focus on decreasing nitrate leaching, as well as preserving forests, specifically rainforests (Selin 2019).

There are also technological options for carbon sequestration in two categories: abiotic and biotic sequestration (Lal 2008). The abiotic options are based on both chemical and physical processes, and biotic is referred to as carbon sequestration in oceans and geological structures (Lal 2008). The mining sector seeks to capture carbon through the method of washing and carbonation of certain metallic components found in mine tailings to produce stable carbonates (Valderrama et al. 2019).

As large deposits of tailings have physical and chemical characteristics that are harmful to health, a good management method is important and is a major challenge for all countries. To address this issue, environmental regulations have been developed. In the case of Mexico, LGEEPA is in charge of establishing norms to prevent and control the effects derived from exploration and exploitation of natural resources, in addition to establishing sites for mining tailings. On the contrary, standard NOM-151-SEMARNAT-2003 issues specifications to characterise these sites and also monitors tailings' tanks. Hence, the mining site owner must register a monitoring programme that approves the performance of protection actions (Ministerio de Minería 2018).

To ensure the stability of tailings or mine tailings, Mexican legislation has set indications as project, construction, and operation criteria. These consider climatic, soil, geotechnical, hydrological, and ecosystem aspects, among others. In addition, factors such as dam storage of tailings, volumetric weight, and surface run-off management must be considered under the civil works design manual of the Federal Electricity Commission. There are also criteria for construction and post-operation actions (Ministerio de Minería 2018).

4 Discussion and Conclusion

This study allowed us to understand that the importance of mining in Mexico dates back to centuries. It was found that the first established law regulating mining activities was enacted in 1892, and the success of this industry depended on the establishment and actions of such federal laws. It was observed that on the one hand, these laws evolved for the interests of decision-makers, on the other, for development of the country, although too many concessions were granted to foreign companies.

Although mining has had a favourable impact on the country's industrial and economic development, the negative effects on the society and the environment are also notable. In reference to impact on the environment and in response to what led to this research work, it was found that since the establishment of regulations in 1992 concerning mining activities, with laws such as LGEEPA and the General Law on Climate Change, it was not until 2014 that environmental care was legally integrated into the Mining Law, and it should be noted that there were NOMs that focussed on soil and air pollution levels.

The mining industry is one of the sectors that produces maximum carbon emissions; however, most of these emissions do not come from extraction, but rather from transporting the products obtained. That is why, all countries developing this activity are facing new challenges. Therefore, new actions, for example, fixing price of the carbon used to transport certain minerals, are required (CEPAL 2019). An avenue to curtail national emissions within this industry could involve participation in international emissions trading systems. There is a robust legal basis that encourages actions to reduce and mitigate climate change. That is why, the mining sector should take advantage of playing a role in achieving carbon sequestration, because CO_2 can be extracted in a natural manner and encapsulated through the rocks commonly found in tailings, thus generating new carbonate minerals. This way, carbon can be permanently stored. In addition, chemical weathering of mine tailings and hydroxide minerals can react with CO_2 to generate magnesium carbonate, which is a cement-like product with potential to store carbon for thousands of years (Williams 2020).

As observed, Mexico has too many mining centres that are relevant for national and international investments; so, these should be more strictly regulated in terms of the environment. One proposal would be that future reforms to the Mining Law and LGEEPA should include specifications on the process of sequestration of carbon emitted during mining processes.

Acknowledgements Instituto Politécnico Nacional, Project SIP: 20220811.

References

Alegría, M.A., León Balderrama, J.I. and Cabanillas López, R. (2021). La minería sonorense, pieza clave en el cumplimiento de los 17 Objetivos del Desarrollo Sostenible (ODS) en la agenda 2030. *Revista Vértice Universitario*, **23(91):** 14–23. https://doi.org/10.36792/rvu.vi91.33

Ali, S.H., Giurco, D., Arndt, N., Nickless, E., Brown, G., Demetriades, A. and Yakovleva, N. (2017). Mineral supply for sustainable development requires resource governance. *Nature*, **543(7645):** 367–372. https://doi.org/10.1038/nature21359

Armendáriz-Villegas, E.J., Covarrubias-García, M., de los A., Troyo-Diéguez, E., Lagunez, E., Arreola-Lizárraga, A. et al. (2015). Metal mining and natural protected areas in Mexico: Geographic overlaps and environmental implications. *Environmental Science and Policy*, **48:** 9–19. https://doi.org/10.1016/j.envsci.2014.12.016

Auditoría Superior de la Federación (ASF). (2015). Evaluación Núm. 1579 Política Pública de Minería. ASF, Mexico

Ávila, C. (August, 2022). Un año crucial para los enlaces entre la industria y el gobierno. *Mexico Business*. Retrieved on 05–09–20233. from https://mexicobusiness.news/mining/news/2022-crucial-year-industry-government-liaisons.

AZO Mining. (April, 2014). Leaching – Mining Fundamentals. Retrieved on March 2022, from: https://www.azomining.com/Article.aspx?ArticleID=1227

Bastidas-Orrego, L.M., Ramírez-Valverde, B., Cesín-Vargas, A., Juárez-Sánchez, J.P., Marínez-Carrera, D. and Vaquera-Huerta, H. (2018). Socioenvironmental conflicts and open-pit mining in the Sierra Norte De Puebla, Mexico. *Textual*, **72:** 35–66. https://doi.org/10.5154/r.textual.2017.72.003

Cámara Minera de México (CAMIMEX). (November, 2020). Importancia de la minería en México. Retrieved on february 2022, from http://www.geomin.com.mx/pdf/panel/litio/ImportanciaMineria_MX_ForoLitio.pdf

Cárdenas, J. (2013). La minería en México: Despojo a la nación. *Revista Mexicana de Derecho Constitucional*, **28:** 35–74.

Carrillo, E. (2022). AMLO publica reforma a la ley minera que nacionaliza el litio. *Forbes*. Available at: https://www.forbes.com.mx/amlo-publica-reforma-a-la-ley-minera-que-nacionaliza-el-litio/

Castillo, K. (2020). Concesiones mineras. *GEOMIMET*. Available at: https://www.revistageomimet.mx/2020/10/concesiones-mineras-en-mexico/

Castro-Soto, G. (November, 2013). La mina nos extermina. *Otros Mundos AC/Chiapas*. Retrieved on March 2022 from: ttps://otrosmundoschiapas.org/la-mina-nos-extermina-manual

Centro de Estudios de las Finanzas Públicas (CEFP). (2018). El Sector Minero y su Importancia en Las Finanzas Públicas de México 2007–2017. *Centro de Estudios de Las Finanzas Públicas (CEFP). Poder Legislativo Federal. Cámara de Diputados*, Brasília, DF-Brazil.

Centro Mexicano de Derecho Ambiental (CEMDA). (2019). Minería a cielo abierto. Available at: https://www.cemda.org.mx/mineria-a-cielo-abierto/#_ftn7

CEPAL. (2019). Minería Para un Futuro Bajo en Carbono. Naciones Unidas, Santiago, Chile.

Chávez, V. and Usla, H. (2022). A detalle: ¿De qué trata la ley minera de AMLO que busca nacionalizar el litio? El Financiero. Available at: https://www.elfinanciero.com.mx/economia/2022/04/18/a-detalle-de-que-trata-la-ley-minera-de-amlo-que-busca-nacionalizar-el-litio/

Concha, E. (2017). Minería global contemporánea. *Ola Financiera*, **10(27):** 81–116. https://doi.org/10.22201/fe.18701442e.2017.27.61009

Contreras, C. (1996). Las ordenanzas de minería de 1783. La polémica entre Gamboa y el Tribunal de Minería. *Historia*, **36:** 39–54. Available at: https://revistas.inah.gob.mx/index.php/historias/article/view/13922

Cortez-Lugo, M., Álamo-Hernández, U. and Hernández-Bonilla, D. (July, 2021). Los efectos de la actividad minera minera en la salud de los Mexicanos. *El Cuerpo Ante La Crisis Ambiental*. Retrieved on March 2022 from: https://discapacidades.nexos.com.mx/los-efectos-de-la-actividad-minera-en-la-salud-de-los-mexicanos/

de la Fuente López, A. and Olivera Villa, B. (2017). Las actividades extractivas en México: Estado actual. *In:* Anuario 2016, Instituto Juvenilia (ISSUU), Palo Alto, CA.

Domínguez and Scheffler. (January, 2022). Mexico's position addressing climate change: A tax policy analysis. Retrieved on February 2022 from: https://www.internationaltaxreview.com/article/b1w654v9t9xcz5/mexicos-position-addressing-climate-change-a-tax-policy-analysis

Environmental Law Institute. (January, 2000). Prevención de la contaminación minera: propuesta de un marco común para las Américas. Retrieved on February 2022 from: https://www.eli.org/sites/default/files/eli-pubs/d10_02s.pdf

Garside, M. (2022a). Mining, metals & minerals. Available at: 2022 Global List of Leading Mining Companies Based on Revenue. https://www.statista.com/statistics/272707/ranking-of-top-10-mining-companies-based-on-revenue/

Garside, M. (2022b). Mining, metals & minerals. Available at: BHP Revenue 2005–2022. https://www.statista.com/statistics/264182/bhp-billiton-revenue-since-2004/

Gaytán, E. and Benita, F. (2014). La industria minera en México: Patrones de desempeño y determinantes de eficiencia (The mining industry in Mexico: Performance patterns and determinants of effiency). *Lecturas de Economía*, **80:** 71.

Gobierno de México. (2015). Industria minera. Situación macroeconómica. Retrieved on March 2022 from: https://www.gob.mx/cms/uploads/attachment/file/119064/Sector_Industria_Mineria.pdf

Gómez Godoy, C. and Peláez Padilla, J. (2020). Minería en México: Despojo, contaminación, conflictos y movilización. Retrieved on March 2022 from: https://www.gob.mx/cms/uploads/attachment/file/581439/Miner_a_en_M_xico__Despojo_contaminaci_n_conflictos_y_movilizaci_n

Hernández, K. and Navarro, C. (September, 2020). Crisis ambiental. Así Se ve La Minería En México. Retrieved on March 2022 from: https://medioambiente.nexos.com.mx/asi-se-ve-la-mineria-en-mexico/

Hernández, Ú., Martínez, A., García, L.A., Fini, D. and Alcántara, R. (October, 2018). Minería y Privilegios. *OXFAM México*. Retrieved on February 2022 from: https://www.oxfammexico.org/sites/default/files/Mineri%CC%81a%20y%20privilegios%20-%20WEB_0.pdf

Herrera, I. and Parra, A. (2018). La actividad minera y el lugar que ocupa en la economía mexicana del siglo XIX. Available at: http://journals.openedition.org/caravelle/3575

Johansson, N., Krook, J., Eklund, M. and Berglund, B. (2013). An integrated review of concepts and initiatives for mining the technosphere: Towards a new taxonomy. *Journal of Cleaner Production*, **55:** 35–44. https://doi.org/10.1016/j.jclepro.2012.04.007

Lal, R. (2008). Carbon sequestration. *Philosophical Transactions of the Royal Society B: Biological Sciences*, **363(1492):** 815–830. https://doi.org/10.1098/rstb.2007.2185

Llano, M. (2016). Concesiones de Agua Para Las Mineras. Fundación Heinrich Böll (Heinrich Böll Foundation), Berlin, Germany.

López, L. and Ruvalcaba, J. L. (2015). El oro de Tenochtitlan: la colección arqueológica del Proyecto Templo Mayor. *Estudios de Cultura Náhuatl*, **49:** 7–57. Retrieved from http://www.scielo.org.mx/scielo.php?script=sci_arttext&pid=S0071-16752015000100002&lng=es&tlng=es

Martínez-Castilla, Z. (2003). Guías prácticas para situaciones específicas: Manejo de riesgos y preparación para respuestas a emergencias mineras. Comisión Económica para América Latina y el Caribe (CEPAL) División de Recursos Naturales e Infraestructura. https://repositorio.cepal.org/handle/11362/2197

Meinert, L.D., Robinson, G.R. and Nassar, N.T. (2016). Mineral resources: Reserves, peak production and the future. *Resources*, **5(1):** 14; xxx. https://doi.org/10.3390/resources5010014

KPMG International Cooperative (2013). Country Mining Guide. MiningMexico. Retrieved on March 2022 from: https://assets.kpmg.com/content/dam/kpmg/pdf/2015/05/mexico-mining-%20guide-revised.pdf

Ministerio de Minería. (June, 2018). Estudios de normativas internacionales de diseño, construcción, operación, cierre y post cierre de depósitos de relaves. Retrieved on march 2022 from: https://www.sernageomin.cl/wp-content/uploads/2018/07/NormativasInternacionalesRelaves.pdf

Misra, K.C. (2000). Understanding mineral deposits. Available at: https://doi.org/10.1007/978-94-011-3925-0.

Murray, C. (2020). Increased indigenous displacement within Mexico and its connections to mining operations in Guerrero. Justice in Mexico, Human Rights and Civil Society, San Diego, CA.

Panossian, C. (April, 2019). La cámara Argentina de empresarios mineros presentó un informe sobre el desarrollo de la industria minera, pp. 5–7. Retrieved on february 2022 from: https://www.caem.com.ar/wp-content/uploads/2019/10/NP_CAEM-presento%CC%81-informe-industria-minera-26-04v2.pdf

Parejo, C. and Parejo, J. (2012). La Minería Metálica En El Mundo. El Caso Particular De Extremadura. La Agricultura y La Ganadería Extremas. Caja de Ahorros de Badajoz, Madrid, Spain, pp. 103–118.

Parlamaint of Australia. (2015). Australia's relationship with Mexico, submission 36. Available at: https://www.aph.gov.au/DocumentStore.ashx?id=4fde5d9c-a8aa-45e3-aa15-a163064d4946&subId=405007.

Procuraduría Federal de Protección al Ambiente. (2016). Norma Oficial Mexicana NOM-159-SEMARNAT-2011 Que establece los requisitos de protección ambiental de los sistemas de lixiviación de cobre. Available at: https://www.gob.mx/profepa/documentos/norma-oficial-mexicana-nom-159-semarnat-2011

Procuraduría Federal de Protección al Ambiente. (2019). Manifestación de impacto ambiental. Available at: https://www.gob.mx/profepa/articulos/manifestacion-de-impacto-ambiental-mia

R. del Bosque, E. (2021). Leyes y reglamentos Mineros México. Available at: https://iclg.com/practice-areas/mining-laws-and-regulations/mexico

Ramírez, É. (May, 2017). Mineras acaparan, contaminan y sobreexplotan el agua de México. *Investigación*. Retrieved on March 2022 from: https://contralinea.com.mx/mineras-acaparan-contaminan-y-sobrexplotan-el-agua-de-mexico/

Saavedra, R. and Hernández, P.Y. (2015). La Exploración. Mi México Es Minero, Tercera edición, pp. 49–52. Available at: https://www.geomin.com.mx/publicaciones/pub3_MEXICO%20MINERO%203%20edicion.pdf

Secretaría de Economía. (2022). Portafolio de proyectos mineros mexicanos. Retrieved on march 2022 from: https://www.economia.gob.mx/files/gobmx/mineria/portafolio_de_proyectos.pdf

Secretaría de Energía. (2015). Secretaría de energía. Available at: CCUS: Tecnología de Captura, Uso y Almacenamiento de Bióxido de Carbono https://www.gob.mx/sener/articulos/ccus-tecnologia-de-captura-uso-y-almacenamiento-de-bioxido-de-carbono

Secretaría de Gobernación (SEGOB). (2004). Norma Oficial Mexicana NOM-141-SEMARNAT-2003 Que establece el procedimiento para caracterizar los jales, así como las especificaciones y criterios para la caracterización y preparación del sitio, proyecto, construcción, operación y postoperación de presas de jales. Diario Oficial de La Federación (SEGOB), Mexico City. Available at: https://www.dof.gob.mx/nota_detalle.php?codigo=661988&fecha=13/09/2004

Secretaría de Gobernación (SEGOB). (2007). Norma Oficial Mexicana NOM-147-SEMARNAT/SSA1–2004 Que establece criterios para determinar las concentraciones de remediación de suelos contaminados por arsénico, bario, berilio, cadmio, cromo hexavalente, mercurio, níquel, plata, plomo, selenio, talio y/o vanadio. Available at: http://www.dof.gob.mx/nota_detalle.php?codigo=4964569&fecha=02/03/2007#:~:text=NORMA%20Oficial%20Mexicana%20NOM%2D147,%2C%20talio%20y%2Fo%20vanadio.

Secretaría de Medio Ambiente y Recursos Naturales. (2003). Norma oficial Mexicana NOM-001-SEMARNAT-1996, que establece los límites máximos permisibles de contaminantes en las descargas de aguas residuales en aguas y bienes nacionales. Available at: https://www.profepa.gob.mx/innovaportal/file/3290/1/nom-001-semarnat-1996.pdf

Secretaría de Medio Ambiente y Recursos Naturales. (2004). Norma oficial Mexicana NOM-141-SEMARNAT-2003, que establece el procedimiento para caracterizar los jales, así como las especificaciones y criterios para la caracterización y preparación del sitio, proyecto, construcción, operación y post-operación de presas de jales. Available at: http://www.ordenjuridico.gob.mx/Federal/PE/APF/APC/SEMARNAT/Normas/Oficiales/NOM-141-SEMARNAT-2003

Selin, N.E. (2019) Carbon sequestration. Britannica Encyclopedia. Available at: https://www.britannica.com/technology/carbon-sequestration.

Servicio Geológico Mexicano (SGM). (2017). Anuario estadístico de la minería Mexicana, 2016. Available at: https://www.sgm.gob.mx/productos/pdf/Anuario_2016_Edicion_2017.pdf

Servicio Geológico Mexicano (SGM). (2020). Consulta los panoramas mineros estatales. Available at: https://www.gob.mx/sgm/articulos/consulta-los-panoramas-mineros-estatales

Servicio Geológico Mexicano (SGM). (2021). Anuario estadístico de la minería Mexicana, 2020. Retrieved on february 2022 from https://www.sgm.gob.mx/productos/pdf/Anuario_2020_Edicion_2021.pdf

Singh. U. (2013). Carbon capture and storage: An effective way to mitigate global warming. *Current Science*, **105:** 7.

Statista. (2021). Global leading mining companies by revenue 2021 ranking. Available at: https://www.statista.com/statistics/272707/ranking-of-top-10-mining-companies -based-on-revenue/

Stevens, R.A. (2010). Mineral exploration and mining. *British Columbia Institute of Technology.*

Suárez, K. (2021). López Obrador eleva el pulso contra las empresas mineras en México. El País. Available at: https://elpais.com/mexico/2021-04-03/lopez-obrador-eleva-el-pulso-contra-las-empresas-mineras-en-mexico.html

Touchpoint Marketing. (2020). El rol de la minería en la transición a una economía baja en carbono. Available at: https://www.thermofisher.com/blog/cienciaacelerada/mineria/el-rol-de-la-mineria-en-la-transicion-a-una-economia-baja-en-carbono/

USGS. (December, 2008). Carbon sequestration to mitigate climate change. Retrieved on february 2022 from https://pubs.usgs.gov/fs/2008/3097/pdf/CarbonFS.pdf

Valderrama, Campusano and Espindola. (2019). Chilean mining: Capture, transport, and storage of carbon dioxide in tailings by ionic liquids and mineral carbonation. *Información Tecnológica,* **30(5):** 357–372. https://doi.org/10.4067/S0718-07642019000500357

Vela Almeida, D., León, M. and Lewinsohn, J.L. (2021). *Indicadores de sostenibilidad en la minería metálica.* Retrieved on march 2022 from https://repositorio.cepal.org/server/api/core/bitstreams/fdf1b9d2-3d51-46d4-b0ca-f68af0a05ef7/content

Violatti, C. (2014). Stone age. World History Encyclopedia. Available at: https://www.worldhistory.org/Stone_Age/

Williams. (2020). Mining technologies could capture 'billions of tonnes of CO_2 per year, says UBC prof. Available at: https://www.mining.com/mining-technologies-could-capture-billions-of-tonnes-of-co2-per-year-says-ubc-prof/

Wilton, J. (January, 2015). El "peor desastre ambiental de los tiempos modernos" en México. Retrieved on March 2022 from: https://newint.org/features/web-exclusive/2015/01/08/mexico-environment-disaster

World Bank. (2016). Development of a regulatory framework for carbon capture, utilization and storage in Mexico. Final Report. Retrieved on February 2022 from: https://documents1.worldbank.org/curated/pt/630261467089960257/pdf/MergedCCUS-Mexico-Regulatory-Framework-Revised-Final-Report.pdf

Zheng, X., Wang, R., Wood, R., Wang, C. and Hertwich, E.G. (2018). High sensitivity of metal footprint to national GDP in part explained by capital formation. *Nature Geoscience*, **11(4):** 269–273. https://doi.org/10.1038/s41561-018-0091-y

Energy-Generation Processes Using CO_2

Francisco Javier Sanchez-Ruiz, Maria Elena Tavera Cortes, and Genoveva Rosano Ortega

1 Introduction

In recent decades, different institutions have monitored global temperatures since the end of the nineteenth century to establish appropriate strategies for mitigating pollutants. In the 1980s, the Goddard Institute for Space Studies (GISS) realised that global warming was underway. The most recent data show an increase in global temperature of at least 0.8 °C since 1951, and the growth continues in the current century, reaching an approximate increase of 1.2 °C (Nocito and Dibenedetto 2020; Talapaneni et al. 2020). In parallel, there has been a substantial increase in the emission of greenhouse gases (GHG) during the last century of human activity, with carbon dioxide being the most crucial gas and the highest concentration, which has modified the very dynamics of the atmosphere (Sullivan et al. 2021). CO_2 emissions into the atmosphere continue to increase steadily each year, reaching a current value of 46.65 billion metric tonnes of CO_2 in 2020. As a result, the atmospheric concentration of CO_2 has risen from the pre-industrial value of 280–424 ppm in 2022 (Han and Ho 2021); this has generated an area of opportunity for the capture of this pollutant for energy cogeneration from the implementation of capture and cogeneration processes, seeking energy sustainability, as well as its mitigation. The Intergovernmental Panel on Climate Change (IPCC) has reported that excessive GHG emissions into the atmosphere, mainly CO_2, are the primary source of modern global climate change and the resulting global warming of the earth (Gabrielli et al. 2020). There is political opposition to the hypothetical human origin of modern climate change. However, the scientific community maintains almost unanimous agreement on this issue. Figure 1 shows the data obtained by the Mauna Loa observatory in Hawaii, USA, where it shows the linear trend of generation and accumulation of CO_2 in the atmosphere, implying a climatic and atmospheric problem (Hospital-Benito et al. 2020).

Despite all efforts to minimise CO_2 emissions, tens of billions of tonnes are emitted yearly. Therefore, the IPCC has proposed carbon recovery as a necessary technique to suppress an increase in the atmospheric concentration of CO_2 (Hu et al. 2019). Traditional carbon recovery typically addresses CO_2 emissions from significant stationary sources, such as power plants. For example, cement factories, oil refineries, and the steel industry are proposed for capturing CO_2 and implementing it in the generation and cogeneration of energy (Fig. 2).

1.1 Pre-combustion

Pre-combustion techniques are applied when you have a natural or synthetic gas (carbon monoxide [CO] and hydrogen [H_2]) process which can use many methods to produce feedstocks suitable for power generation, and CO_2 captures, such as steam reforming (Reaction 1), pulverised coal (integrated gasification combined cycle [IGCC]), coke, and residual oil gasification (Reaction 2). The water–gas shift reaction, a reaction between CO and water, produces a gas stream primarily consisting of CO_2 and hydrogen (Reaction 3). Removal of CO_2 allows the high-purity H_2 stream for power generation. The CO_2 concentration is 15–60% in these cases, so CO_2 recovery is relatively easy. On the other hand, the first fuel preparation steps for coal gasification and steam reforming are demanding and expensive (Breyer et al. 2019):

$$C_xH_y + H_2O \rightarrow xCO + \left(x + \frac{y}{2}\right)H_2$$

$$C_xH_y + O_2 \rightarrow xCO + \left(\frac{y}{2}\right)H_2$$

F. J. Sanchez-Ruiz · G. R. Ortega
Universidad Popular Autónoma del Estado de Puebla, Puebla, Mexico

M. E. T. Cortes (✉)
Instituto Politécnico Nacional—UPIICSA, Ciudad de México, Mexico
e-mail: mtavera@ipn.mx

J. Marrugo-Negrete et al. (eds.), *Remediation and Carbon Capture in Mining Sector*, https://doi.org/10.1007/978-3-031-87883-1_7

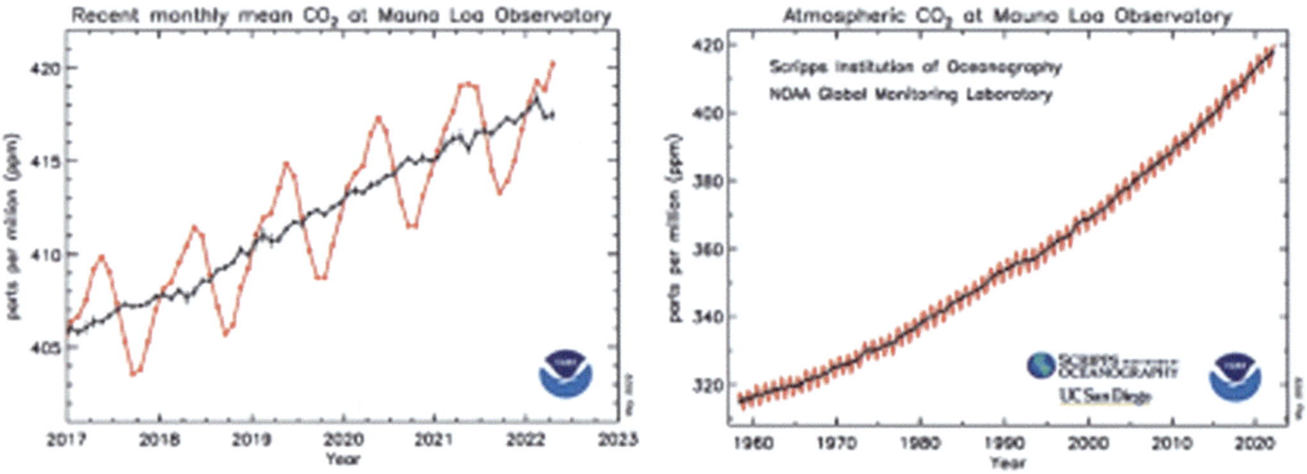

Fig. 1 Monitoring of generation and concentration of CO_2 in the atmosphere

Fig. 2 CO_2-capture processes

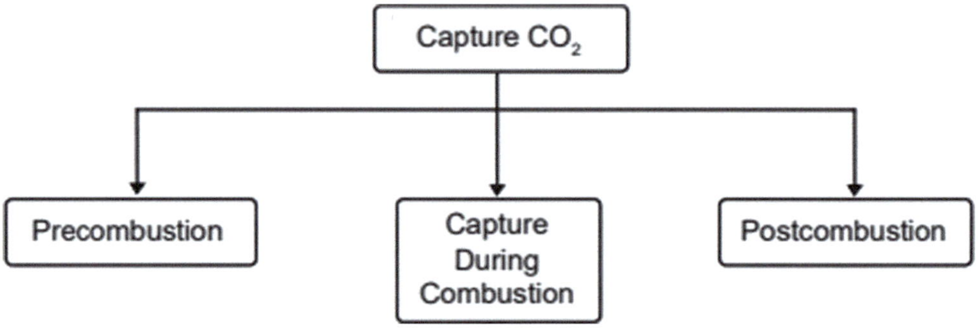

$$CO_2 + H_2O \rightarrow CO_2 + H_2$$

$$C_xH_y + H_2O, \text{Üí } xCO + \left(x + \frac{y}{2}\right)H_2, \tag{1}$$

$$C_xH_y + \frac{x}{2}O_2, \text{Üí } xCO + \left(\frac{y}{2}\right)H_2, \tag{2}$$

$$CO + H_2O, \text{Üí } xCO_2 + H_2, \tag{3}$$

1.2 Capture During Combustion

Two main processes used for capture during combustion are oxy-firing and carbonation-calination. Oxy-fuel processes discharge with pure oxygen instead of air—a part of combustion gases recirculated in approximately 30–40% of such processes. The flue gas volume is reduced by using oxygen instead of air, and the CO_2 concentration continuously increases (up to 90%). However, these technologies are still not highly favoured due to the high expenses of air separation required to obtain pure oxygen (Ghiat and Al-Ansari 2021; Jung et al. 2019).

1.3 Post-combustion

It can also capture CO_2 from flue gases released after the combustion process and can use various technologies, but it is optional to modify existing combustion installations. In general, the combustion gas must have a previous treatment, such as desulfurization, to obtain a gas with greater processing capacity; this prior treatment process is part of most countries' regulations or environmental laws. The main disadvantage of this approach is the reduction in CO_2 concentration, which ranges from 4 to 6% by volume for natural gas-combined cycling processes to 15–17% by volume for a coal-fired plant (Sen et al. 2020; Yamada 2021; Tsiotsias et al. 2020). Impurities, such as SO_x and NO_x, are also a problem for many separation processes; for example, because of their reactivity towards amines, which are the compounds commonly used for post-combustion CO_2 capture, amine-based separation processes require flue gas pretreatment to significantly reduce concentrations of these strongly acidic gases (Ochedi et al. 2020).

Post-combustion CO_2 separation processes on an industrial scale for energy generation have been based mainly on liquid amine and polyamine absorption techniques. Aqueous solutions of monoethanolamine (MEA), diethanolamine

(DEA), and methyl-di-ethanol-amine (MDEA) have been successful in capturing CO_2 on an industrial scale (Feron et al. 2020). The problems or disadvantages of this process are related to the corrosive properties of liquid amines (Tao et al. 2020), amine losses, energy consumption in the amine-regeneration processes, and polyamines (Adamu et al. 2020). What has caused new technologies to be sought and generated for CO_2 capture and may constitute these operations on cryogenic technologies, adsorption, and functional organic membranes (Hidalgo and Martín-Marroquín 2020).

Carbonation-calcination uses the carbonation reaction between quicklime (calcium oxide [CaO]) and CO_2 to produce limestone (calcium carbonate [$CaCO_3$]) (Shirmohammadi et al. 2020). $CaCO_3$ is introduced into a calcination reactor, where the reverse process occurs. Calcination produces regenerated quicklime and a CO_2-enriched gas stream. The main advantages are the low cost of the initial limestone used and its reactivity against CO_2 and SO_2. Therefore, no desulfurization step is required before carbonation-calcination. However, the high temperature used during carbonation (650 °C) and calcination (900 °C) reactions and material wear are significant economic drawbacks—the carbonation step with flue gas in a CO_2 post-combustion.

2 Energy Production

One of the most implemented processes for separating CO_2 from gas streams is the application of aqueous amine solutions. This approach is practiced commercially to remove CO_2 from CO_2-rich natural gas streams and is now implemented on submarines to purify breathing air. The most significant number of reports describing investigations of adsorbents have focused on the use of solid-supported amine materials. A chemical reaction occurs between CO_2 and amines in supported amine materials, creating solid bonds and allowing significant pickups even at low CO_2 partial pressures (Bos et al. 2020; Danaci et al. 2020). Consequently, the heat of sorption and selectivity towards CO_2 are higher than for sorbents. This feature makes amine-based organic–inorganic hybrid materials highly suitable for direct air capture (Fig. 3).

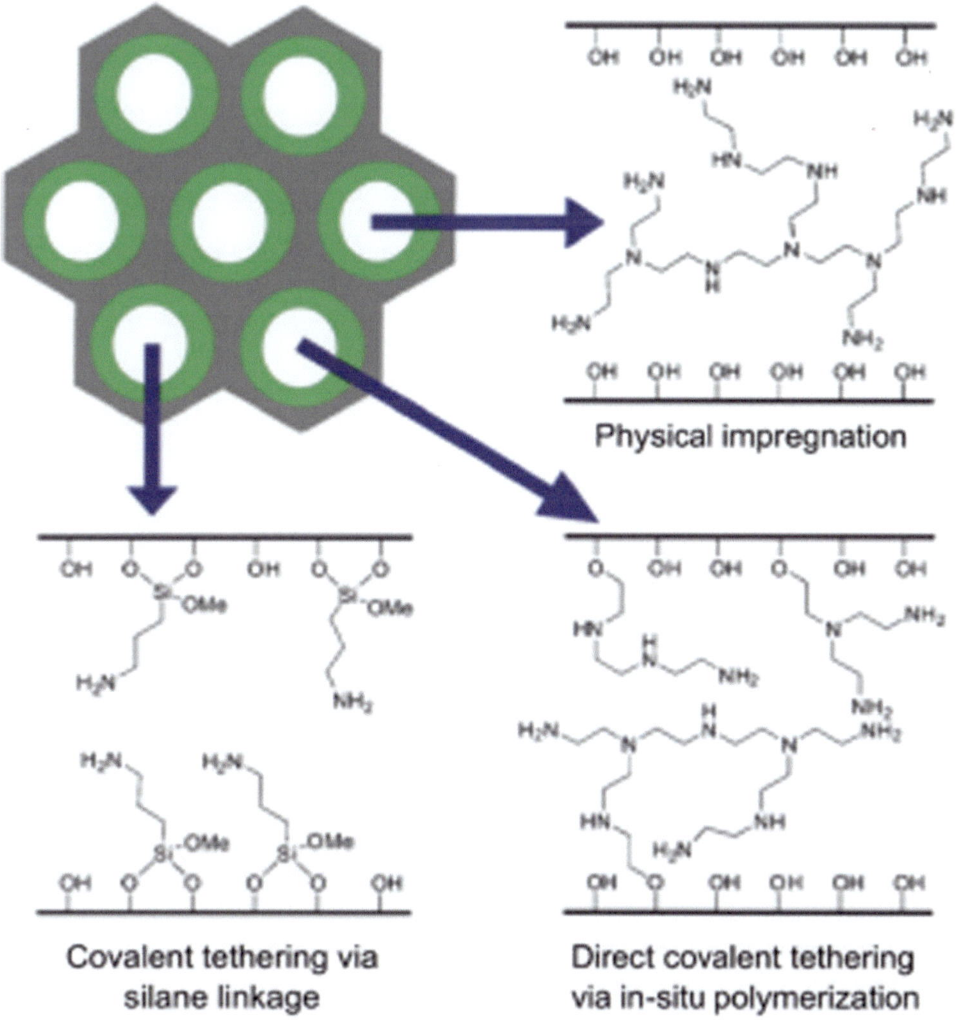

Fig. 3 Duplicate CO_2 adsoiption from amaines

Table 1 Combinations of amines and adsorption capacity (Zhang et al. 2019)

Support	Amine	T (°C)	CO_2 amine sorbent	Capacity sorbent (mm CO_2/g sorbent)	Adsorption capacity (mmol of CO_2/g of sorbent)	Amine efficiency (mmol of CO_2/mmol of NHx)
Business Silica	PEI	25	400	10.5	2.36	0.22
	PEI + AP	25	400	10.7	2.26	0.21
	PEI + TP	25	400	10.5	2.19	0.21
MCF	PAA	25	400	7.2	0.86	0.12
	PEI	25	400	10.7	1.74	0.16
	PEI-lnH	25	400	11.4	1.05	0.09
Zr-SBA-15	PEI	25	400	8.3	0.85	0.1
γ-alumina	PEI	25	400	11.2	1.74	0.16
SBA-15	PEI	25	400	9.2	1.05	0.11
Business Silica	TEPA	35	400	10.1	2.5	0.25
Smoked silica	PEI-H	25	420	33.0	1.18	
		25	420	33.0	1.77	
		25	420	55.0	1.71	
		25	420	55.0	1.41	
SBA-15	PEI	75	400		0.51	
γ-alumina	PEI	30	400	8.5	1.96	0.25
Smoked silica	PEI-H + PEG	75	360	9.8	0.68	0.07
SBA-15	PEI + PEG200	30	400	5.8	0.79	0.14
Smoked silica	PEI-ln	25	400	55.0	2.34	–

* PEI poly (ethyleneimine); PEI-H poly (ethyleneimine) high; PEI-lnH poly (ethyleneimine longhigh; TEPA tetraethylenepentamine; PAAA poly (acrylamine); PE polyethylene; TA tetramine; G200 compund 200; PEI-ln poly (ethyleneimine) long; SBA-15 subtract based acid; MCF flourinated mesocarbon

The impregnation process, in which the chemical species are deposited in the pores or on the surface of solid support, is a versatile process that allows almost any species to be loaded onto a given material because chemical reactions generally do not occur. Therefore, various amine-containing species load onto the porous oxide support. However, because temperature swing adsorption (TSA) is often applied directly to air acquisition, volatility is primarily compared to limiting adsorbent amines' loss during degradation and regeneration of adsorbent. Low-grade amine species, low molecular weight branched poly(ethyleneimine) (PEI), have a high amine group density. It is very stable in TSA or vacuum swing adsorption (VSA), making it an essential choice for many researchers to absorb CO_2. In addition, Zhang et al. (2019) investigated the utility of small amines, such as MEA, used in solution processes with solid adsorbents. Unfortunately, its low boiling point (usually 140–245 °C) caused considerable evaporation. Other linear amines, such as tetraethylenepentamine (TEPA) and pentaethylenehexamine (PEHA), typically also show some amine leaching or loss, although not as low in molecular weight (Table 1).

A comparison of process efficiency was made according to the adsorption capacity. Figure 4 shows the potential application for power generation.

2.1 Design of CO_2-Capture Process

The plant consists of a natural gas-fired gas turbine (GT) producing a nominal 160 MWh electricity. It has a Heat Recovery Steam Generator (HRSG) to recover waste heat from GT exhaust and generate superheated steam that powers three 27-MWh steam turbines to produce additional power. The nominal total generation capacity of the plant is 243 MWh (Fig. 5).

Likewise, different processes establish a high efficiency of energy production by capturing CO_2 using liquid amines.

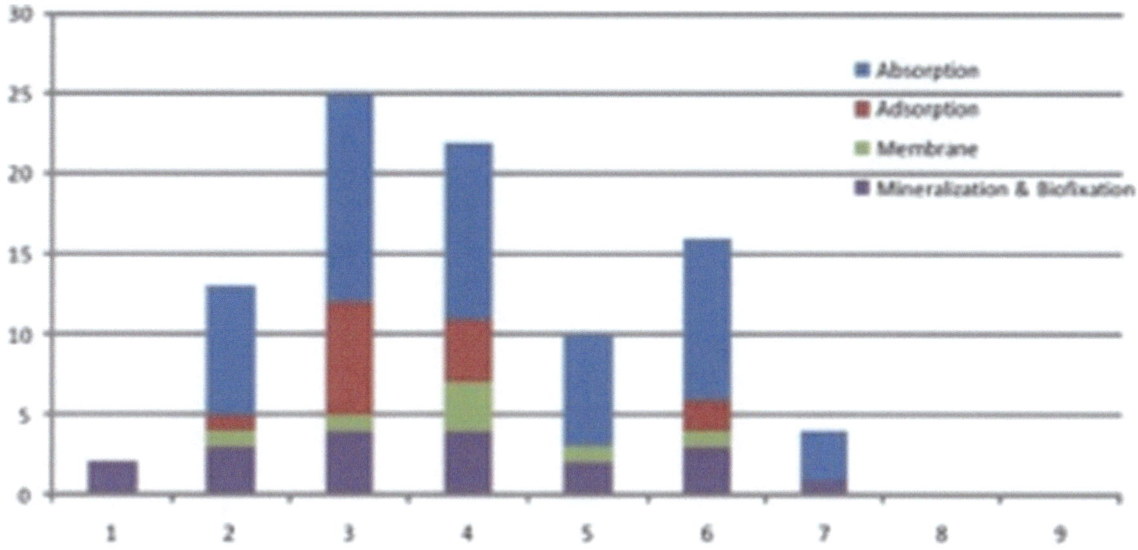

Fig. 4 Capacity by technology

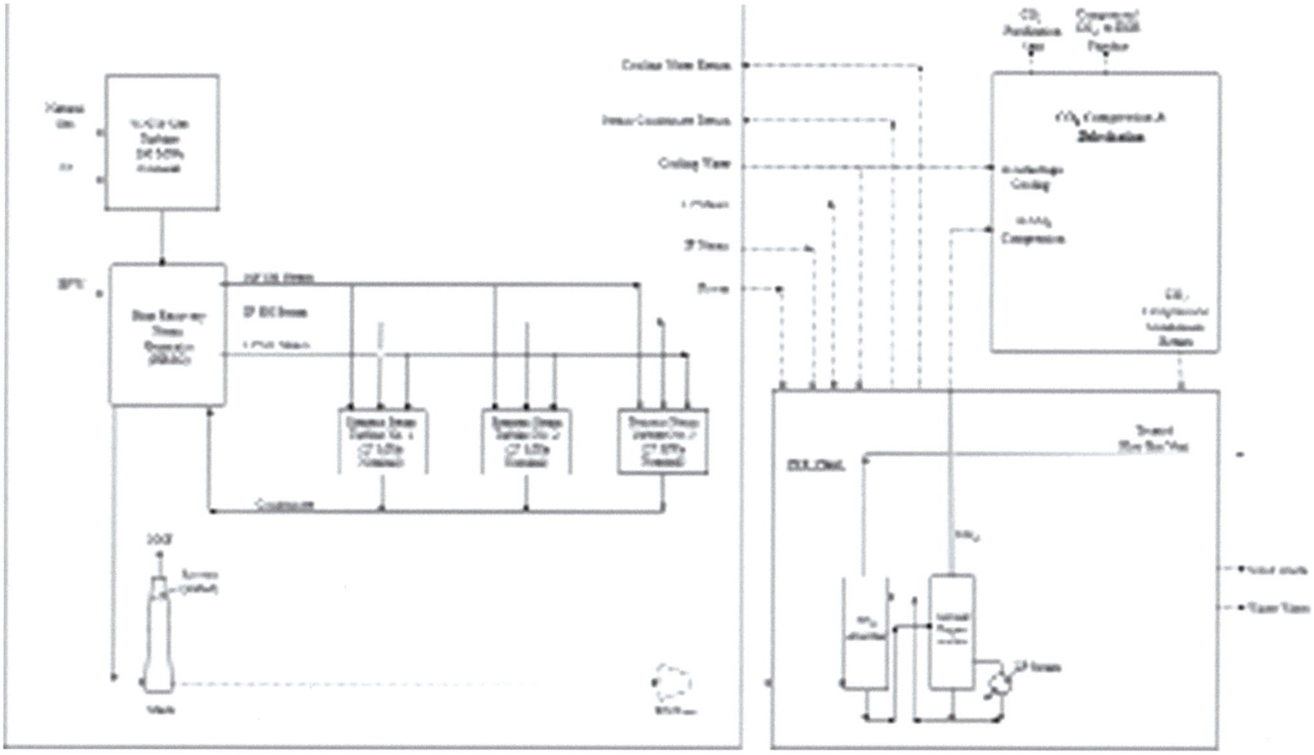

Fig. 5 General diagram of energy production

2.1.1 Advanced Flow Scheme Process

The Advanced Flow Scheme (AFS) process configuration is a technology developed for CO_2 capture that provides minimal energy consumption, leading to reduced operating costs. The carbon-capture facility uses mono and polyamine technology. This type of process adapts to power plants that have fossil fuels as a source of supply, and its implementation is in petrochemical production and carbon-capture systems CO_2 from the air (Mikulčić et al. 2019; George et al. 2021; Wilberforce et al. 2021).

The process comprises two packed columns with monoethanolamine and polyamines (MEA and PEHA). In column C1, the CO_2 treatment and three adsorption stages separate the dome area, corresponding to the release of fuel gas regenerated without CO_2, which is recirculated into the process. The bottom phase to a second C2 column in which

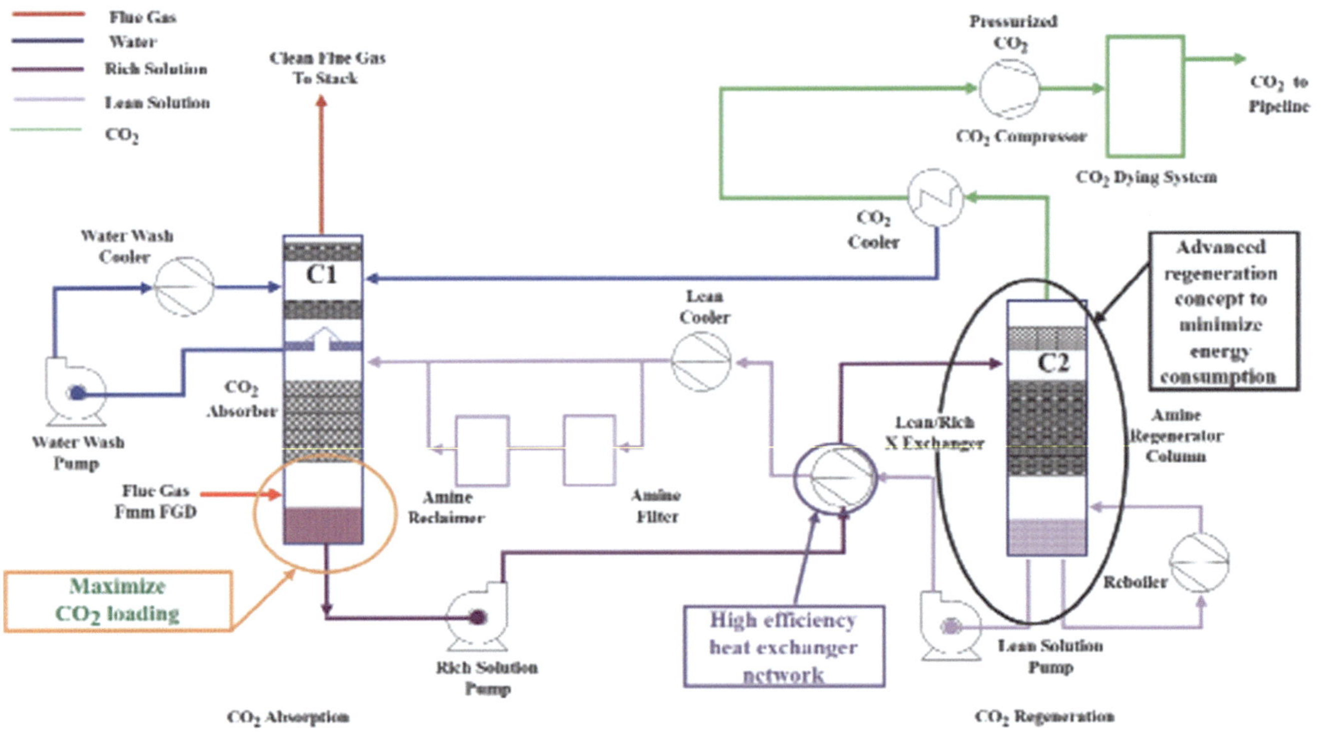

Fig. 6 Capture processing of CO_2

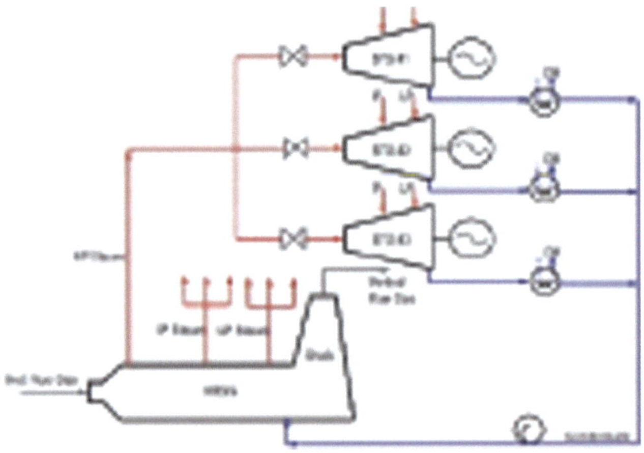

Fig. 7 Power generation employing turbines

CO_2 is fractionated from amines (Fig. 6). Once the CO_2 is captured in a storage container, it feeds a system of energy-cogeneration turbines (Fig. 7); the CO_2 removal and capture capacity is 80–90% efficiency, with a flow of 295,000 Nm^3/h as a cold gas.

2.1.2 Blue Process

It is called a blue process because of CO_2-capture materials, which have the characteristic of presenting a faint blue colour before gas capture, subsequently presenting an intense colour but with the property of being regenerated very easily (Sifat and Haseli 2019).

This process uses two split-wall reactive columns, which have higher efficiency than counter-flow adsorption columns. This implies that the CO_2 capture and treatment capacity in this process is between 280,000 and 400,000 Nm^3/h, and the gas obtained is transported directly to the energy-cogeneration system, continuously generating approximately 2.5 MW-h/day with minimum flow and up to 600 MW-h/day with maximum flow.

Figure 8 shows the process schematically, where both columns are coupled thermally, the divided wall of separation columns is in design bases, and the energy and thermal efficiency in this process makes it more feasible than the one mentioned above.

2.1.3 Intensified Process

Compared to the two technologies mentioned above, the intensified CO_2-capture processes are more commonly used in the industry because of their capture capacity and low cost (Stančin et al. 2020; Abdin and Mérida 2019). These processes use coupled and modified separation processes to capture CO_2, intensified due to the minimum amount it implements for energy production and separation of components. The efficiency of this process is 90–95% capture and separation of CO_2 as fuel in energy production, which reaches a production of 250–800 MW-h/day.

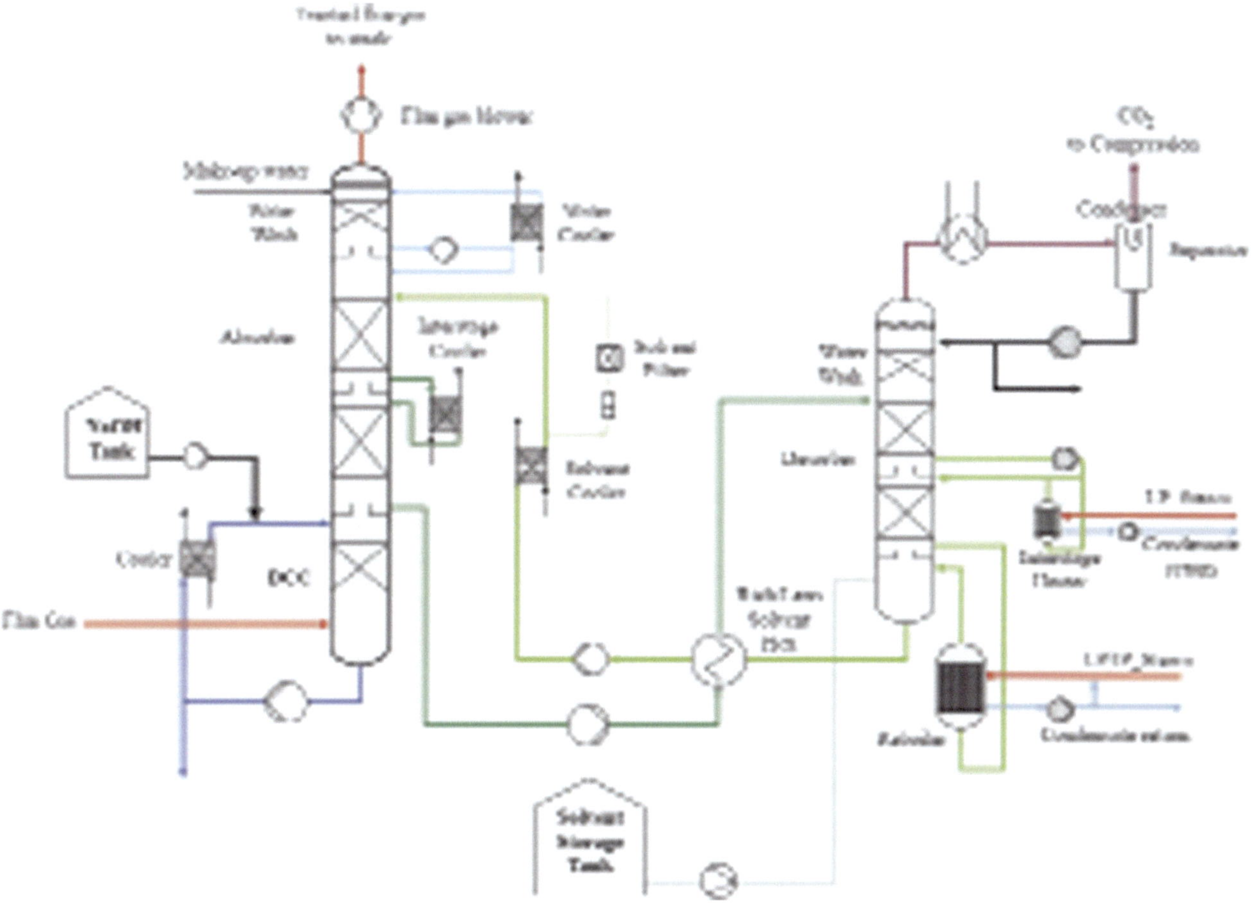

Fig. 8 Capture and separation process of CO_2 using divided wall columns

Figure 9 schematically shows the intensified process for capturing and producing energy through CO_2. The sequence is first through an adsorption column, which could be a packed column or fluidised bed. The charged gas passes through a simple separation column, but it can also be a reactive column; that is, it has a double coupled process, a reactor integrated into a separation column, although this last option considerably increases the cost of the process. However, it can achieve an efficiency of up to 98% separation.

2.1.4 Process Using Hydroxides

The CO_2-capture process implementing bases, such as sodium hydroxide, is a commonly implemented process, mainly because of the cost of reagents. One of the disadvantages of the process is the cost of materials used for constructing separation columns; this is primarily due to the oxidation capacity of the reagents used, as CO_2 has a high oxidation potential. The energy generation through this process is low due to the CO_2-capture capacity; the efficiency is 75–80%, with an energy production of 10–200 MW-h/day. Several successful cases of the implementation of this type of technology are reported in different countries, making it one of the most feasible processes to implement, but with certain reservations because of its energy production capacity as well as mitigation of carbon capture (Guo et al. 2019; Capros et al. 2019).

Figure 10 shows the process schematically. It is observed that only two pieces of equipment are present: an evaporation column, where a mixture of gas-NaOH enters in a hydroxide CO_2 is recirculated when the dome stream contains CO_2 with a minimum concentration of NaOH can capture CO_2 when evaporating the dome stream has mainly CO_2 with a minimum concentration of NaOH. The bottom stream is entirely NaOH, which is mixed before the column. Gas from the dome is condensed and fed to the turbine system for energy generation. Each process uses the same number of turbines to compare the methods' efficiencies and the energy production capacity using CO_2 as a fluid of work.

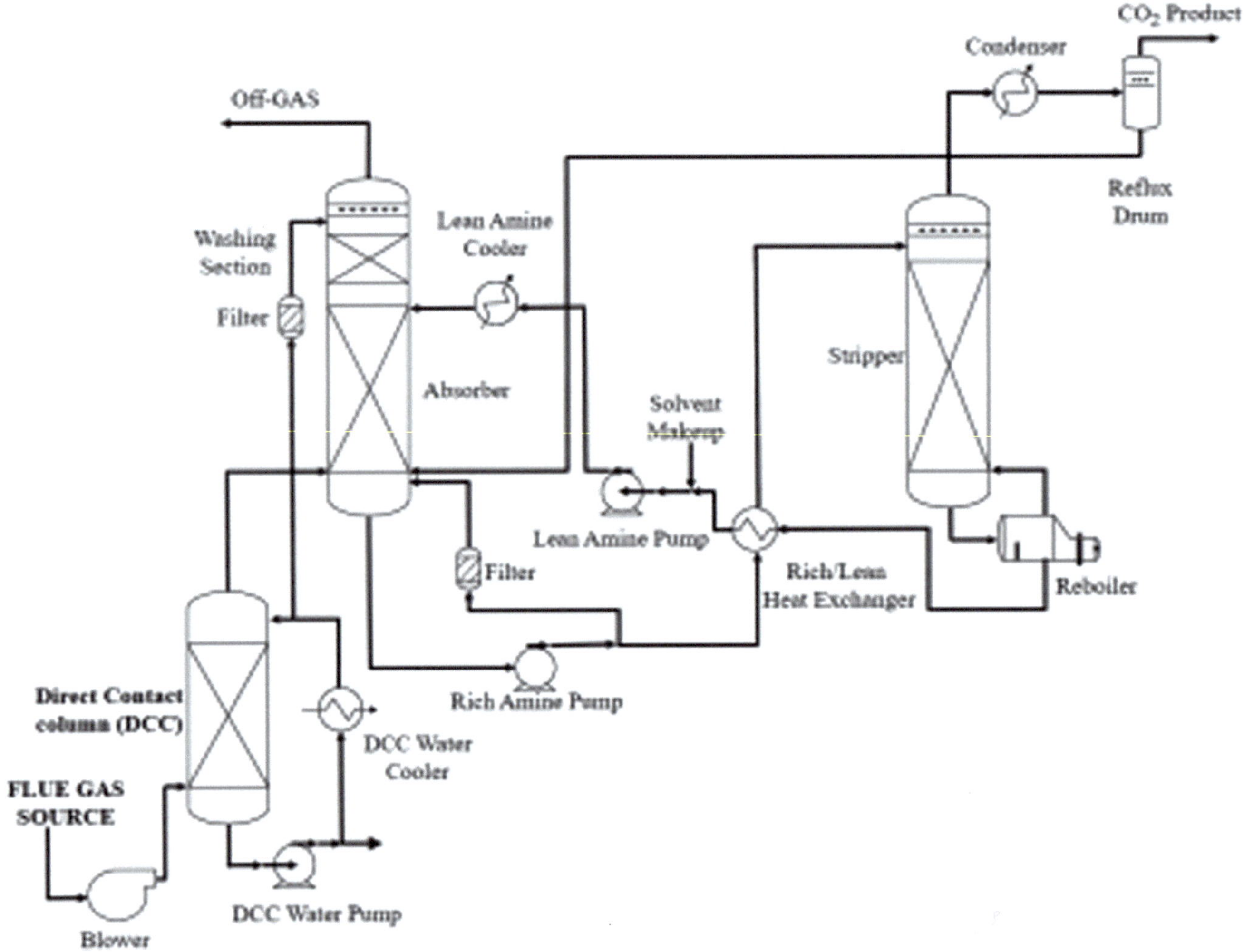

Fig. 9 Intensified process

2.1.5 Capture Process by Multi-stage Cooling

Capturing CO_2 through coupled cooling systems is one of the most straightforward processes with good efficiency. This process consists of three stages of sequential cooling. The first stage is cooling and capturing water in a cooling column; the second stage consists of a cooling tower where refrigerants, such as ammonium, R134–R136 in some cases, can be water at a temperature close to 5 °C (278 K) used as a working fluid. Compared to refrigerant fluids, the efficiency is 50%, which increases significantly to 60–70%. One of the limitations of the use of refrigerant fluids is the capacity of reintegration into the process (Gambhir et al. 2019), which does not happen when water is used as a working fluid; approximately 95% of the water flow that enters the process is recovered, which makes the process a sustainable process which does not happen when water is used as a working fluid, approximately 95% of the water flow that enters the process. Compared to other methods proposed in this section, the power-generation capacity through this process is considered low; approximately 20–150 MW-h/day power is generated (Fig. 11).

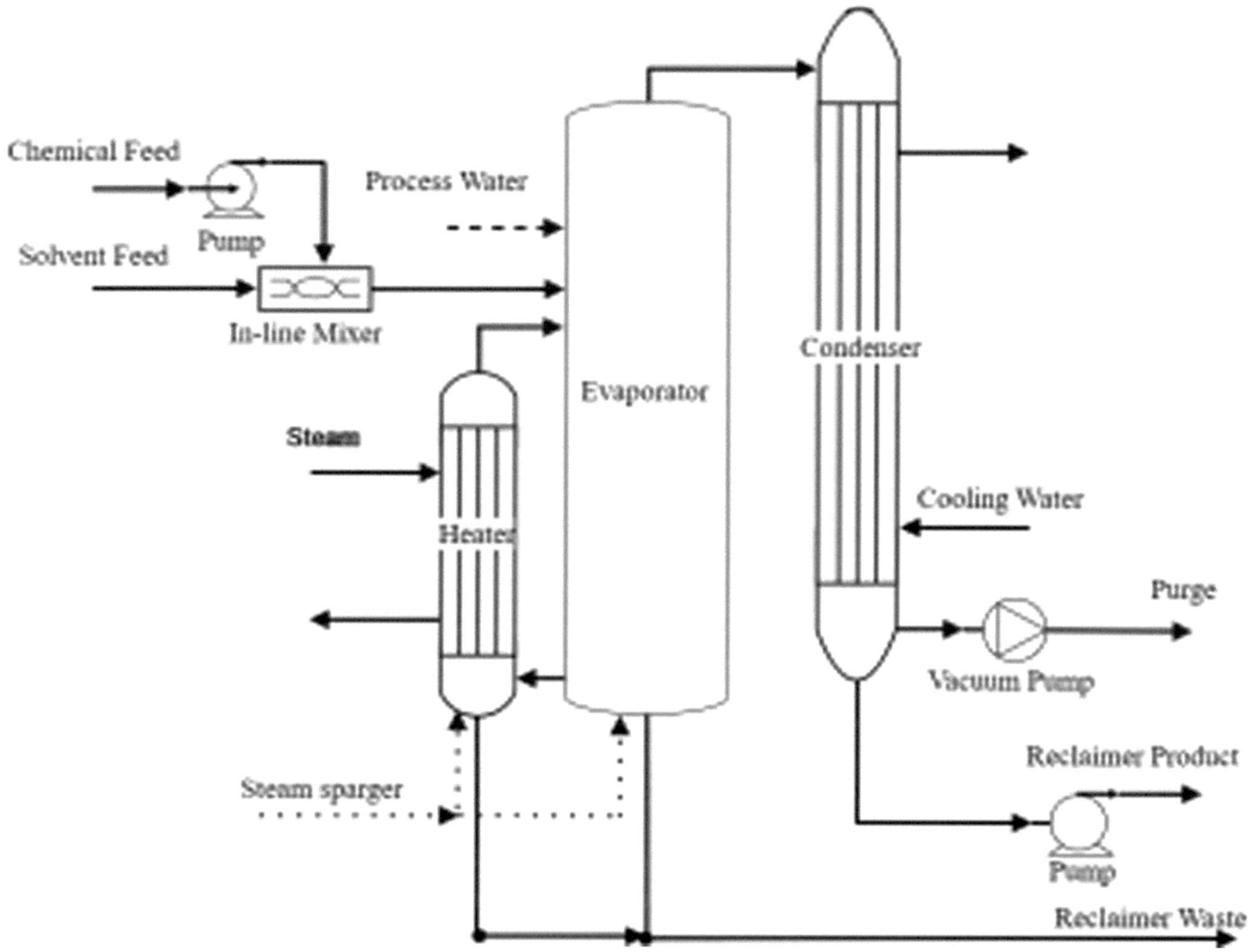

Fig. 10 Duplicate CO_2 capture process using NaoH

2.1.6 Capture Process in Geothermal and Petrochemical Plants

The capture of CO_2 in geothermal and petrochemical processes is different from those in which the capture is pre-combustion or post-combustion. The capture of CO_2 in this process is in the preceding or subsequent stages, which can be in the adjusted high or low enthalpy steam cooling. The condenser inlet and energy recovery are for geothermal processes. To have greater efficiency in capturing CO_2, fluidised media, comprising SO_x and NO_x, are used, which cause displacement reactions to the existing process to capture CO_2 and is more efficient. In petrochemical processes, the capture is exercised by implementing the same type of fluids; comparing the different adsorption or CO_2-capture processes has shown that they have a greater efficiency than traditional energy generation processes, specifically those related to geothermal processes. However, the fractionating column is at the points of maximum CO_2 production, the medium and low molecular weight hydrocarbon stages. In this process, the CO_2-capture efficiency is 70–90%, with a flow of 700,000 Nm^3/day and an energy production of 300–900 MW-h/day. Figure 12 schematically shows the capture process (Chehrazi and Moghadas 2022).

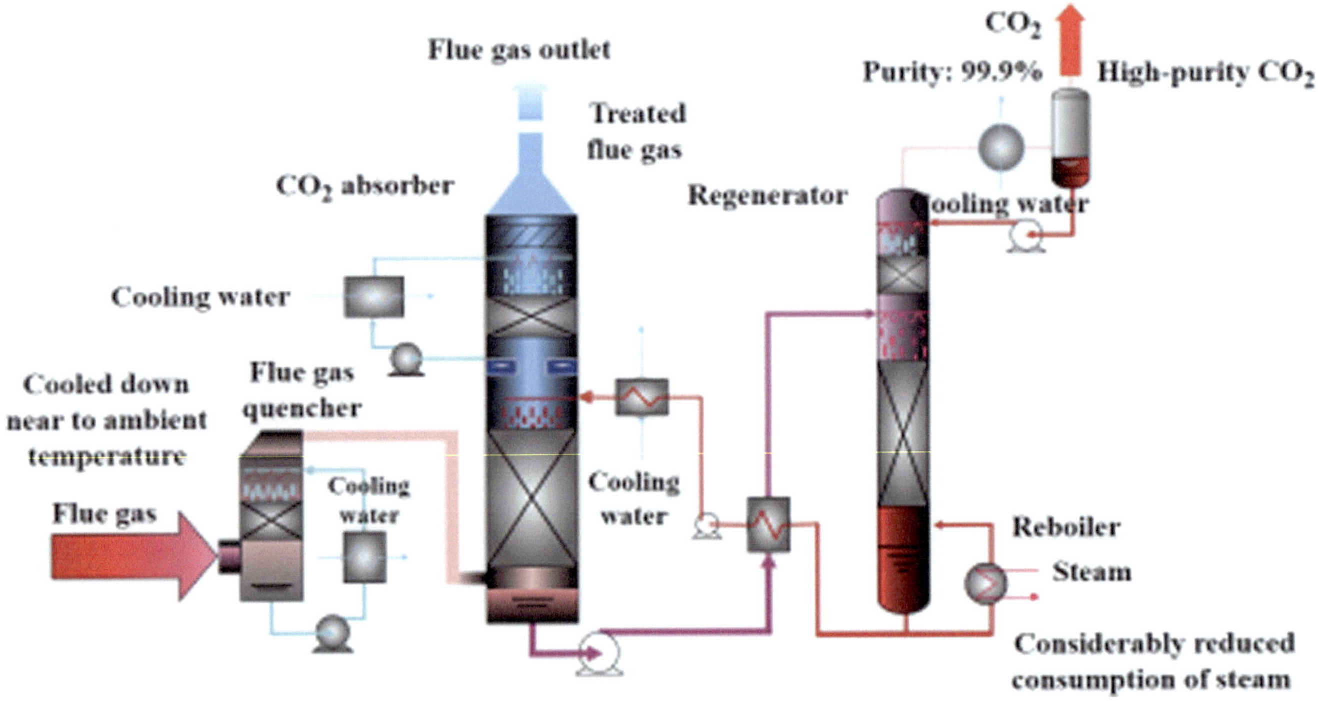

Fig. 11 The multi-stage cooling capture process

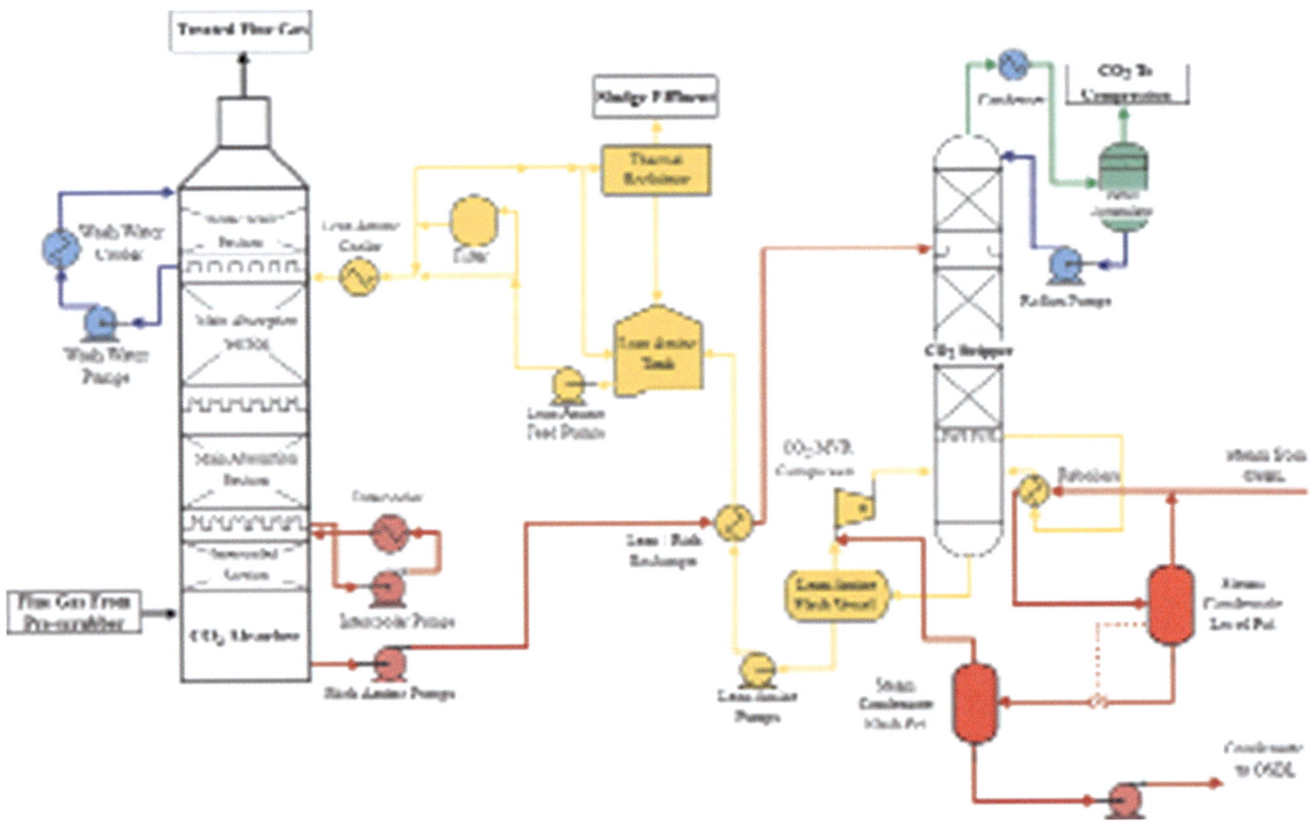

Fig. 12 Petrochemical and geothermal capture process

3 Conclusion

Using technologies based on the capture of CO_2 is a viable alternative. From the decarbonization perspective, the use of adsorption mechanisms with liquids, such as amines, provides efficiencies and high energy production, and these technologies include an output of 10–900 MW-h/day. In the process with energy production potential, the efficiency is 50–95%. It indicates that they are processed with yields even when their input flow is low, compared to other methods, such as energy production through hydroelectric plants, which require high volumes of water to produce 900 MW-h/day. Currently, no technologies are considered zero emissions, mainly in energy production, especially because each technology can present a certain amount of CO_2 emission that can be used for energy generation through its capture and transformation. In addition, if there is no zero-emission energy, 20 and 30 kg CO_2/MW-h are produced.

Water is a process of capturing CO_2; it is feasible from the perspective of energy production but unfeasible from the environmental point of view because of the high volumes of water used. However, there is an aspect where saline water could be used to capture CO_2, but the high equipment cost could make using this type of fluid for CO_2-capture unfeasible. The capture of CO_2 for energy production and decarbonization is a measure that should be implemented gradually. In most industrialised countries, this results in a 10–15% reduction of CO_2, which decreases ground temperature.

References

Abdin, Z. and Merida, W. (2019). Hybrid energy systems for off-grid power supply and hydrogen production based on renewable energy: A techno-economic analysis. *Energy Conversion and Management*, **196:** 1068–1079.

Adamu, A., Russo-Abegão, F. and Boodhoo, K. (2020). Process intensification technologies for CO_2 capture and conversion – A review. *BMC Chemical Engineering*, **2(1):** 1–18.

Bos, M.J., Kersten, S.R.A. and Brilman, D.W.F. (2020). Wind power to methanol: Renewable methanol production using electricity, electrolysis of water, and CO_2 air capture. *Applied Energy*, **264:** 114672.

Breyer, C., Fasihi, M., Bajamundi, C. and Creutzig, F. (2019). Direct air capture of CO_2: A key technology for ambitious climate change mitigation. *Joule*, **3(9):** 2053–2057.

Capros, P., Zazias, G., Evangelopoulou, S., Kannavou, M., Fotiou, T., Siskos, P.,and Sakellaris, K. (2019). Energy-system modelling of the EU strategy towards climate-neutrality. *Energy Policy*, **134:** 110960 https://doi.org/10.1016/j.enpol.2019.110960

Chehrazi, M. and Moghadas, B.K. (2022). A review on CO_2 capture with chilled ammonia and CO_2 utilization in urea plants. *Journal of CO_2 Utilization*, **61:** 102030.

Danaci, D., Bui, M., MacDowell, N. and Petit, C. (2020). Exploring the limits of adsorption-based CO_2 capture using MOFs with PVSA–from molecular design to process economics. *Molecular Systems Design & Engineering*, **5(1):** 212–231.

Feron, P.H., Cousins, A., Jiang, K., Zhai, R. and Garcia, M. (2020). An update of the benchmark post-combustion CO_2-capture technology. *Fuel*, **273:** 117776.

Gabrielli, P., Gazzani, M. and Mazzotti, M. (2020). The role of carbon capture and utilization, carbon capture and storage, and biomass to enable a net-zero-CO_2 emissions chemical industry. *Industrial & Engineering Chemistry Research*, **59(15):** 7033–7045.

Gambhir, A., Rogelj, J., Luderer, G., Few, S. and Napp, T. (2019). Energy system changes in 1.5 C, well below 2 C and 2 C scenarios. *Energy Strategy Reviews*, **23:** 69–80.

George, A., Shen, B., Craven, M., Wang, Y., Kang, D., Wu, C. and Tu, X. (2021). A review of non-thermal plasma technology: A novel solution for CO_2 conversion and utilization. *Renewable and Sustainable Energy Reviews,* **135:** 109702.

Ghiat, I. and Al-Ansari, T. (2021). Carbon capture and utilization review as a CO_2 abatement opportunity within the EWF nexus. *Journal of CO_2 Utilization*, **45:** 101432.

Guo, H., Li, C., Shi, X., Li, H. and Shen, S. (2019). Nonaqueous amine-based absorbers for energy-efficient CO_2 capture. *Applied Energy*, **239:** 725–734.

Han, Y. and Ho, W.W. (2021). Polymeric membranes for CO_2 separation and capture. *Journal of Membrane Science*, **628:** 119244.

Hidalgo, D. and Martín-Marroquín, J.M. (2020). Power-to-methane, coupling CO_2 capture with fuel production: An overview. *Renewable and Sustainable Energy Reviews*, **132:** 110057.

Hospital-Benito, D., Lemus, J., Moya, C., Santiago, R. and Palomar, J. (2020). Process analysis overview of ionic liquids on CO_2 chemical capture. *Chemical Engineering Journal,* **390:** 124509.

Hu, Z., Wang, Y., Shah, B.B. and Zhao, D. (2019). CO_2 capture in metal-organic framework adsorbents: An engineering perspective. *Advanced Sustainable Systems*, **3(1):** 1800080.

Jung, S., Park, Y.K. and Kwon, E.E. (2019). Strategic use of biochar for CO_2 capture and sequestration. *Journal of CO_2 Utilization*, **32:** 128–139.

Mikulčić, H., Skov, I.R., Dominković, D.F., Alwi, S.R.W., Manan, Z.A., Tan, R. and Wang, X. (2019). Flexible carbon capture and utilization technologies in future energy systems and the utilization pathways of captured CO_2. *Renewable and Sustainable Energy Reviews*, **114:** 109338.

Nocito, F. and Dibenedetto, A. (2020). Atmospheric CO_2 mitigation technologies: Carbon capture utilization and storage. *Current Opinion in Green and Sustainable Chemistry*, **21:** 34–43.

Ochedi, F.O., Liu, Y. and Adewuyi, Y.G. (2020). State-of-the-art review on the capture of CO_2 using adsorbents prepared from waste materials. *Process Safety and Environmental Protection*, **139:** 1–25.

Sen, R., Goeppert, A., Kar, S. and Prakash, G.S. (2020). Hydroxide-based integrated CO_2 capture from air and conversion to methanol. *Journal of the American Chemical Society*, **142(10):** 4544–4549.

Shirmohammadi, R., Aslani, A., Ghasempour, R. and Romeo, L.M. (2020). CO_2 utilization via integration of an industrial post-combustion capture process with a urea plant: Process modeling and sensitivity analysis. *Processes*, **8(9):** 1144.

Sifat, N.S. and Haseli, Y. (2019). A critical review of CO_2 capture technologies and prospects for clean power generation. *Energies*, **12(21):** 4143.

Stančin, H., Mikulčić, H., Wang, X. and Duić, N. (2020). A review on alternative fuels in future energy systems. *Renewable and Sustainable Energy Reviews*, **128:** 109927.

Sullivan, I., Goryachev, A., Digdaya, I.A., Li, X., Atwater, H.A., Vermaas, D.A. and Xiang, C. (2021). Coupling electrochemical CO_2 conversion with CO_2 capture. *Nature Catalysis*, **4(11):** 952–958.

Talapaneni, S.N., Singh, G., Kim, I.Y., Al Bahily, K., Al-Muhtaseb, A.A.H., Karakoti, A.S., and Vinu, A. (2020). Nanostructured carbon nitrides for CO_2 capture and conversion. *Advanced Materials*, **32(18):** 1904635.

Tao, L., Huang, J., Dastan, D., Wang, T., Li, J., Yin, X. and Wang, Q. (2020). CO_2 capture and separation on charge-modulated calcite. *Applied Surface Science*, **530:** 147265.

Tsiotsias, A.I., Charisiou, N.D., Yentekakis, I.V. and Goula, M.A. (2020). The role of alkali and alkaline earth metals in the CO_2 methanation reaction and the combined capture and methanation of CO_2. *Catalysts*, **10(7):** 812.

Wilberforce, T., Olabi, A.G., Sayed, E.T., Elsaid, K. and Abdelkareem, M.A. (2021). Progress in carbon capture technologies. *Science of the Total Environment*, **761:** 143203.

Yamada, H. (2021). Amine-based capture of CO_2 for utilization and storage. *Polymer Journal*, **53(1):** 93–102.

Zhang, S., Shen, Y., Wang, L., Chen, J. and Lu, Y. (2019). Phase change solvents for post-combustion CO_2 capture: Principle, advances, and challenges. *Applied Energy*, **239:** 876–897.

Techno-Economic Evaluation of Carbon Dioxide Capture Technologies: A State-of-the-Art Review of Chemical Absorption Processes

Foster Kofi Ayittey and Agus Saptoro

1 Introduction

Carbon emissions are one of the world's problems that requires immediate, sustainable solutions. Carbon dioxide (CO_2) capture and storage (CCS) technologies are being explored extensively. Among other options for minimising CO_2 emissions from centralised emission points, they are deemed highly promising, including power plants, chemical production facilities, and other industrial processes (Jilvero et al. 2014a, b). Despite the expanding significance of renewable energy, CCS technologies are especially pertinent given the rising global utilisation of fossil fuels (Suman 2021; Mutezo and Mulopo 2021; Deshmukh et al. 2021; Li et al. 2020; Kim 2021). Over the past few decades, extensive research projects have been conducted to upgrade these technologies, improving system performance and affordability (Li et al. 2016a, b).

While further study and development are necessary to meet the criteria for industrial-scale applications in the energy sector (first-generation processes), certain fundamental CCS technology components have already attained advanced technical maturity. These fundamental elements comprise less harmful environmental effects from the use of scrubbing agents, enhanced absorbent conversion efficiency, long-term safe storage, lower capital costs, and the use of CO_2 for energy generation and material production (Kuckshinrichs and Hake 2016). The commercial-scale implementation of CCS technologies continues to face growing scepticism from various segments of society worldwide despite the intense national and international research efforts in these areas, some of which are already focusing on second-generation processes. An integrated technological assessment is urgently required to evaluate and address the technical, economic, environmental, and social aspects of CCS technologies in this setting (Kuckshinrichs and Hake 2016). The technical performance and affordability of CCS technologies continue to be major research topics and obstacles to their full implementation, so works focusing on these technical and economic analyses are especially important (IEA 2011; Figueroa et al. 2008; Rubin et al. 2012).

The present paper examines the most recent research on the techno-economic evaluation of CO_2 capture systems using chemical absorption as the separation mechanism because these processes continue to be the most sophisticated CCS approach. This article focuses on several capture solvents, including potassium carbonate (K_2CO_3), ammonia (NH3), piperazine (PZ), monoethanolamine (MEA), 2-amino-2-methyl-1-propanol (AMP), and their blended solutions. This paper reviews the techno-economic analyses (TEA) of capture systems applied to other carbon emission sources, such as pulp mills, metallurgical industries, and cement plants. On the other hand, most review publications only consider cost estimates for carbon capture systems for power plants. Hence, this article is anticipated to serve as a comprehensive resource on techno-economic evaluations of chemical absorption-based CCS technologies for the spectrum of audiences interested in this field of research.

2 Research Methods

Every effort was made to access and analyse all relevant publications to capture the state-of-the-art publications on the economic analyses of CCS technologies that employ chemical absorption as the means of separation. To find these relevant articles, a systematic literature survey was conducted through

F. K. Ayittey · A. Saptoro (✉)
Curtin Malaysia Research Institute (CMRI), Curtin University Malaysia, Sarawak, Malaysia
e-mail: agus.saptoro@curtin.edu.my

A. Saptoro
Department of Chemical and Energy Engineering, Curtin University Malaysia, Sarawak, Malaysia

J. Marrugo-Negrete et al. (eds.), *Remediation and Carbon Capture in Mining Sector*, https://doi.org/10.1007/978-3-031-87883-1_8

various sources such as (1) Web of Science, (2) Science Direct, (3) Scopus, (4) Wiley Online Library, (5) Google Scholar, (6) Springer Link, (7) Taylor & Francis, (8) American Chemical Society publications, (9) Emerald, and (10) ProQuest.

The keywords included in the search were 'carbon capture, 'ccs technology', 'carbon capture technology', 'carbon separation', 'chemical absorption', 'CO_2 mitigation', 'CO_2 capture', 'MEA-based capture system', 'NH_3-based capture system', 'DEA-based capture system', 'PZ-based capture system' 'K_2CO_3-based capture system', 'carbon capture cost', along with 'economic analyses', 'economic models', 'economic feasibility', 'cost analyses', 'cost assessment', 'technical performance', 'technical parameters', 'technical evaluation', 'technical assessment', 'cost parameters', 'cost evaluation', 'cost estimates', 'cost models', 'economic evaluation', etc.

A total of 62 relevant study materials were selected among the many that were retrieved on CCS technologies. During the literature selection process, articles were carefully checked for the relevance of data, similarities in information, accuracy of presentations, and data compilation dates to select the most recent and relevant literature. The state-of-the-art information compiled from these studies has been discussed at length and systematically in the following sections.

3 Scope and Background

3.1 Audience for This Work

Several stakeholders in industries, governments, and other organisations or agencies can benefit from the knowledge of the technical performance and cost estimates for CCS technologies due to their potential importance in minimising greenhouse gas (GHG) emissions. These stakeholders may use this knowledge for technology assessments, investment decisions, policy analysis, research and development (R&D) activities, and energy-environmental policymaking, including creating laws and regulations relating to CCS technologies (Rubin 2012).

Hence, this comprehensive review article targets two groups: practitioners of techno-economic analyses and the end-users of techno-economic analysis findings. The first group comprises anyone involved in the design, modelling, costing and analysis of CCS technologies. This group is often found in research and R&D organisations, academia, and industrial establishments. The second group, as described by Rubin (2012), are primarily policymakers in governments and those who make technology decisions in R&D agencies. A key example of this group is funding bodies, which consistently make informed decisions on funding CCS research proposals.

3.2 TEA and Its Applications

Techno-economic analysis (TEA) is a comprehensive approach to assessing the economic feasibility and technical performance of an industrial process or value chain. It is sometimes referred to as techno-economic evaluation or techno-economic assessment. Spreadsheet software or process simulators are typically used to estimate the capital cost, operational cost, and income of a process or technology based on changing technical and financial input parameters (Burk 2018; Wikipedia 2021; NREL 2021; VITO 2021). In addition to analysing the advantages and disadvantages of a process, TEA may entail analysing its risks and uncertainties. Therefore, TEA might generally be used to set goals for research and development, guide budgetary decisions, assess the economic viability of a new process, assess the existing process's optimisation parameters, and quantify risks and uncertainties.

TEA is performed using a techno-economic model, which involves an integrated process and a cost model, combining elements of process design, equipment sizing, process modelling, capital cost estimation, operating and maintenance (O&M) cost estimation, and cash flow analysis. According to Guandalini et al. (2019), a simplified method of cost evaluation for a carbon capture technology requires the following input data:

a. The carbon capture approach used: post-combustion, pre-combustion, and oxy-combustion.
b. The energy consumption of the CO_2 separation method per unit mass of CO_2 captured.
c. The proportion of the electric power consumed during the CO_2 capture process is released to the environment via the plant cooling system.
d. The carbon capture efficiency.
e. Heat duty per unit mass of CO_2 captured, provided by steam from the turbine and the steam extraction pressure.
f. The low-temperature heat recovered from the CO_2 capture process in the steam cycle.
g. The purity of the captured CO_2 and the pressure at which it is delivered from the capture unit.

The TEA for a carbon capture plant alone is much simpler to assess. The comprehensive TEA for a CCS technology could be very complex to execute. This is so because detailed parameters of technical performance and elements of cost evaluation for the proposed storage facility and the transport facilities for the compressed CO_2 are also required for a complete TEA to be performed for a typical CCS technology. Many have analysed the economic estimations for various carbon capture plants over the past decade. Nevertheless, the literature has limited comprehensive works on the techno-economic assessment of transport and storage (T&S) facilities for the captured CO_2.

3.3 Indicative TEA Models

In the works of Van der Spek et al. (2020) and Van der Spek et al. (2014), three types of TEA models were described as the most common in current CCS literature. These are simplified TEA models, intermediate complexity TEA models, and detailed or rigorous TEA models. According to the above authors, the first type of TEA model is used to discover a general understanding of the process's technical performance and economic feasibility. These simplified models, often using top-down cost estimates, are thus derived from simple first principles, and their economic evaluations are mostly based on forecasts of well-established equipment costs.

In an intermediate complex model, the models are often more robust than a simple model but usually are not rigorous. These CCS TEA models use shortcut models for technology descriptions, ignoring part of the physical descriptions such as mass transfer, heat transfer, and chemical kinetics, in addition to partial process design and equipment lists for cost estimations. On the other hand, detailed TEA models are more often based on rigorous physical technology models and comprehensive cost estimation models. Standard examples of such models are variants of the conventional bottom-up engineering-economic cost estimate approach typically adopted by popular technology research giants and organisations such as Electric Power Research Institute (EPRI), the International Energy Agency Greenhouse Gas Program (IEAGHG), the Global Carbon Capture and Storage Institute (GCCSI), and the US Department of Energy's National Energy Technology Laboratory (DOE/NETL) (Rubin 2019; Van der Spek et al. 2020).

A less common cost estimation method for CCS technologies is the hybrid method proposed by Rubin (2019). This approach combines a top-down cost analysis based on learning or experience curves that reflect past cost patterns with the conventional bottom-up cost analysis for a new process. Wene (2008) has provided details on these learning curves, while van Vuuren et al. (2009) and Yang et al. (2020) have conducted studies comparing top-down and bottom-down cost estimate approaches and providing evidence linking the two models.

Van der Spek et al. (2020) stated that the technology readiness level (TRL), data availability for the designated technology and physicochemical models, and the objective of the process being studied might all influence the choice of TEA model type.

3.4 Metrics of CCS Technical Performance and Cost Measures

3.4.1 Measures of Technical Performance

This document is not intended to be an exhaustive collection of technical performance indicators for a CCS system. The effectiveness of the plant's carbon capture or removal, the level of duty assigned to the stripper reboiler, the amount of solvent used, and the amount of heat required for absorbent regeneration are often the primary indicators of a solvent-based capture plant's technical success. Additional crucial factors are the solvent's absorption capacity (expressed in mass transfer rates), solvent loss, heat transfer rates, plant heat loss, CO_2 purity level, absorbent material performance, etc. Technical performance indicators deemed acceptable include high carbon capture rates, low stripper reboiler duties, high solvent regeneration rate, high CO_2 absorption level, low solvent loss, high CO_2 purity, and low heat loss from the capture plant.

Three essential elements are considered to measure the technical feasibility of a subsurface CO_2 storage complex. These are the capacity of the storage site, injectivity level, and containment or integrity level (NETL 2021). A good storage site must have sufficient space (capacity) to contain large volumes of compressed CO_2. This storage capacity is a fraction of the pore volume of permeable and porous sedimentary formations available at the site. According to Cooper (2009) and Cooper et al. (2009), the pore volume available for containment depends mainly on five parameters: area of storage site, formation thickness or amount of porous rock available for storage, rock porosity (φ, % of voids per bulk volume), the density of CO_2, and storage efficiency (defined as the fraction of pore volume saturated by the compressed CO_2 fluid). Whereas the storage efficiency for a given storage site remains relatively constant, the other four parameters could vary considerably. Porosity is a crucial factor that dictates pore volume distribution within the formation. In general, porosity values above 10% in carbonate formations and 15% in clastic formations are preferred. A good injection operation also depends on having a sufficient formation thickness. Deviated or horizontal drilling can also be utilised to increase access to the formation. A minimum need for formation thickness is typically around 20 m, though this depends on the injection volume needed. A significant aspect influencing the site capacity may also be the lateral continuity of high-quality rock blocks.

The injectivity of the storage site, a parameter which determines the formation's permeability, refers to the rate at which carbon can be injected into the subsurface. The permeability of a formation, which measures the resistance to fluid flow through it, is closely related to its porosity. Generally, higher porosity corresponds to improved permeability, affecting fluid flow and injectivity. A rock with high porosity tends to have more interconnected pore spaces, allowing fluid movement. Porous rocks exhibit a wide range of permeability values. These values span from approximately 0.1 milli-Darcy (for very tight rocks) to several Darcies (for highly permeable formations). CO_2 storage sites benefit from high permeability values (typically exceeding 100 mD). This high permeability ensures efficient fluid injection near the wellbore, facilitating rapid access to the pore space. While high permeability is desirable, extreme values can have implications. Very high permeability pathways or conduits may enhance CO_2 migration along concentrated routes. Paradoxically, this can reduce the overall effective storage capacity within the target formation. Hence, researchers emphasise balancing high permeability for injectivity and avoiding excessive migration pathways (Cooper et al. 2009).

A storage site's ability to confine CO_2 securely and permanently within a set volume without a breach from the storage complex is its integrity or containment. An overlaying sealing package that guarantees leak-free fluid containment is necessary for this. The distribution and shape of rocks and the pressure systems that restrict fluid flow below the surface are key factors in containment. One capable sealing package drapes and encloses the reservoir in the most basic geometrical situation, also called a four-way dip closure, limiting flow upward. In reality, practical seal geometries that restrict the movement of CO_2 in the subsurface will vary widely, much like oil and gas fields do. These variations include lateral and vertical reservoir variations and complex ultimate seal geometries, including flow barriers like faults and naturally over-pressured zones. It is advisable to conceptualise a storage complex and a sealing system as distinct entities, each comprising many geological features. This approach acknowledges rock units' inherent variability and dynamic nature over time. The objective is to demonstrate the overall security of the geological containment system (Cooper et al. 2009).

3.4.2 Metrics for Cost Estimates

A range of metrics are used in literature to measure the cost of CCS systems. The commonest metrics consist of CO_2 avoided cost, cost of captured CO_2, cost of CO_2 abated, and the increased cost of electricity (often called levelised cost) (Metz et al. 2005; Rubin 2012). An overview of these metrics is summarised in Table 1.

The cost of CO_2 avoided (CCA) or CO_2 avoidance cost (CAC) equals the carbon tax applied to atmospheric emissions at which the cost of electricity of the reference plant equals that of the plant with CCS (Rao and Rubin 2002). Since emissions to the atmosphere are not avoided unless the captured CO_2 is sequestered, the CCA must include the full cost of CCS. However, this is not always the case. Some studies only assess the cost of the carbon capture (CC) plant without incorporating the costs of T&S, which misrepresents the estimate for the CCA or CAC (Rubin 2012; Metz et al. 2005).

The concept of CO_2 capture cost (CCC) is distinct from CCA in that there is no requirement or assumption that the captured CO_2 should be avoided or sequestered. Thus, the numerical value of the cost of CO_2 captured is always lower than that of CO_2 avoided, mainly because of the added cost involved in the T&S of CO_2 in the cost of CO_2 avoided.

Table 1 Metrics for CCS cost estimates based on Rubin (2012)

Metric	Description/Application	Mathematical formula	Unit
Cost of CO_2 avoided (CCA)	• Compares a plant with CCS to a 'baseline plant' without a CCS system • Quantifies the average cost of avoiding a ton of atmospheric CO_2 emissions		$/($tCO_2$)
Cost of CO_2 captured (CCC)	• Measures the economic viability of a CO_2 capture system relative to a market price for CO_2 as an industrial commodity		$/($tCO_2$)
Cost of CO_2 Reduced (CCR)	• Used to estimate the cost of reducing CO_2 emissions involving multiple emission points rather than a single power plant		$/($tCO_2$)
Cost of electricity (COE)	• Used to estimate the cost of electricity generation for a single power plant or a collection of power plants		$/(MWh)

- tCO_2/MWh = CO_2 mass emission rate to the atmosphere in tons per MWh (based on the net capacity of each power plant).
- $(tCO_2/MWh)_{captured}$ = total mass of CO_2 captured per net MWh for the plant with capture (equal to CO_2 produced minus emitted.
- NPV = net present value cost of the specified scenarios, tCO_2 = CO_2 mass emission rate in tons per year.
- TCC = total capital cost ($), FCF = fixed charge factor (fraction/year), FOM = fixed O&M costs ($/year), VOM = variable non-fuel O&M costs ($/MWh), HR = net power plant heat rate (MJ/MWh), FC = unit fuel cost ($/MJ), CF = plant capacity factor (fraction), 8766 = total hours in an average year, and MW = net plant capacity (MW).

When discussing GHG reduction plans involving several CO_2 emission sources rather than just one power plant, the term 'cost of CO_2 abated or reduced' (CCR) refers to a more general cost estimate. In this context, it denotes the price of switching from one scenario with a set energy mix and a given location to another with reduced CO_2 emissions. Conditions with reduced CO_2 emissions frequently involve adjustments to the mix of power generation in addition to demand-reduction strategies, not just CCS. This concept differs from the CCA in that the latter is only defined for a single plant (Rubin 2012; Clarke et al. 2009, 2008; Riahi et al. 2004).

Whether implicitly or explicitly stated, the cost of electricity (COE) or the levelised cost of electricity (LCOE) is the most common metric used in engineering-economic studies of CCS to quantify the increased cost of electricity for a power plant that installs CCS. By definition, the LCOE represents the uniform revenue generated from the sale of electricity that a power plant must realise to fully recover all capital and operating costs while earning a specified rate of return over the plant life (Rubin 2012; Rubin and Zhai 2012).

4 TEA of CCS Systems for Power Plants

4.1 CCS Integrated into a Coal-Fired Power Plant

The most researched carbon capture and storage (CCS) method is chemical absorption-based post-combustion carbon capture (PCC) in pulverised coal-fired power stations. It is considered one of the most expensive CCS technologies for industrial applications despite its relative maturity; to run at full capacity, it needs substantial financial assistance (Zhang 2015; Stromberg 2008). Desideri et al. (2014) presented average cost estimates indicating that the cost of captured CO_2 for this technology ranges from 44 \$/t CO_2 to 64 \$/t CO_2.

Among other solvents, MEA is considered the benchmark absorbent for PCC processes. The technical, economic, and environmental evaluation of an amine (MEA)-based PCC process for an assumed new 500 MW (gross) coal-fired power plant was conducted by Rao and Rubin (2002). The carbon emission rate for the plant without a CCS facility was reported as 941 gCO_2/kWh (net), with a capture efficiency of 90% indicated for the MEA-based capture plant. Technical and cost estimates were completed using the Integrated Environmental Control Model (IECM) framework.

The primary parameters considered when evaluating the capture plant's technical performance are solvent and energy requirements. The mass flow rate of CO_2 in the incoming flue gas to the absorber system, the concentration of MEA in the lean solvent stream, the CO_2 loading in the lean solvent stream, and the intended carbon removal efficiency of the capture plant were shown to be the critical determinants of the MEA requirement. The thermal energy required for solvent regeneration—a factor that mostly depends on solvent loading—as well as the electrical energy needed for CO_2 compression, solvent circulation, and other system requirements, were identified as heat requirements.

Two main components were included in the process cost model. This includes the operating and maintenance (O&M) costs and the total capital requirement (TCR). The TCR of the plant entails the total of the direct equipment costs and indirect costs that are computed as fractions of the total process facilities cost (PFC), following the guidelines of the Electric Power Research Institute (EPRI) adhered to in the cost estimates (EPRI 1993). The sum of the PFC, engineering costs, engineering and home office, general facilities, process contingency, and project contingency was used to estimate the total plant cost (TPC). The O&M costs, which could be separated into the fixed O&M (FOM) and variable O&M (VOM) costs, were also estimated as fractions of the TPC. Details of these cost estimates are shown in Table 2. The authors explained that the energy costs were included in the model by derating the overall power plant based on the energy requirement calculated, causing the cost per net kilowatt-hour supplied by the plant to increase.

The deterministic results from the TEA model analysis for the power plant without and with a CCS facility are shown in Table 3. From the results presented in Tables 2 and 3, it could be observed that adding a CCS unit increases the TPC cost by \$134 million. Also, due to the parasitic effect of the installed CCS unit, the net power output of the plant was reduced by roughly 29.44% while limiting the net carbon emission rate to 133 gCO_2/kWh as against the pre-capture value of 941 gCO_2/kWh. Energy requirements utilise about 27% of gross plant capacity, mainly for solvent regeneration (49% of the total) and CO_2 output compression (34% of the total). Fan power consumption and solvent circulation account for the remaining share (17%). MEA makeup requirements make up about 10% of the cost of CO_2 capture. The net LCOE for the plant with an installed CCS system was noticed to increase by \$ 47.8 MWh, ultimately resulting in a carbon mitigation cost of \$ 59.1/tCO_2 avoided.

The techno-economic performance of a 14.3 wt% ammonia (NH_3)-based PCC plant (with an attached ammonia abatement cycle) coupled to a 408 MW modern coal-fired power station (with a high electric generation efficiency of 47.2%) in Denmark was examined by Jilvero et al. (2014a, b). According to them, this facility emits 73.6 $kgCO_2/s$ of CO_2, which translates to an estimated 2.1 $MtCO_2$ of emissions annually, assuming the plant runs for 8,000 h a year. Aspen In-Plant Cost Estimator was used to generate cost estimates, and the net present value (NPV)

Table 2 Cost model parameters for MEA-based PCC plant (based on Rao and Rubin 2002)

Capital cost elements	Value	O&M cost elements	Value
Process area costs (nine areas)		Fixed O&M costs (FOM)	
Total process facilities cost	PFC	Total maintenance cost	2.5% TPC
Home office and engineering	7% PFC	Maintenance cost allocated to labour	40% of total maintenance cost
General facilities	10% PFC	Admin & support labour cost	30% of total labour cost
Contingency, project	15% PFC	Operating labour	2 jobs/shift
Contingency, process	5% PFC		
Total plant cost (TPC) = sum of above		Variable O&M costs (VOM)	
Interest during construction	Calculated	Absorbent (MEA)	$1250/t MEA
Royalty expenses	0.5% PFC	Water cost	$0.2/m^3$
Preproduction expenses	1 month VOM & FOM	CO_2 transportation	$0.02/t CO_2/km
Inventory (startup) expenses	0.5% TPC	CO_2 storage/disposal	$5/t CO_2
TCR = sum of above		Solid waste disposal	$175/t waste

Table 3 Deterministic TEA results for a 500 MW plant with and without CCS facility (extracted from Rao and Rubin 2002)

Parameter	Units	Reference plant	With CO_2 control
Net plant capacity	MW (net)	462	326
CO_2 emission rate	g CO_2/kWh (net)	941	133
CO_2 sequestered	106 tCO_2/yr		2.58
Cost of electricity	$/MWh (net)	49.2	97.0
CO_2 mitigation cost	$/t CO_2 avoided		59.1

economic analysis method was applied to evaluate operating and capital expenditures (OPEX and CAPEX, respectively). The technical analysis shows that at a CO_2 capture rate of 66.2 kg/s (i.e. 90% capture efficiency), the electrical efficiency of the power plant was reported to decrease by 9.2%, corresponding to electricity generation of 341.8 MW, as compared to the reference capacity of 408 MW. Out of the 9.2% reduction in electrical efficiency, the loss of steam alone accounted for 8.1%, with the remaining 1.1% attributed to pump and compression operations. The specific solvent regeneration energy requirement for the NH_3-based capture process was projected to be 3.16 MJ/kgCO_2. With regards to the economic assessment of the same PCC plant, the CAPEX for a full-scale capture process was estimated at 230 M€, with possible cost reduction options to lower the value to 175 M€. The OPEX was assessed to be 66.5 M€/year, which corresponds to a specific capture cost of 35 €/tCO_2. With potential cost-saving options suggested by the authors to lower the OPEX to 59.4 M€/year, the specific capture cost could be reduced to 31.3 €/tCO_2.

About 32% of China's total CO_2 emissions come from the country's energy sector, primarily from burning fossil fuels. The country's substantial reliance on coal for electricity supply is unlikely to change significantly shortly, as coal-fired power plants already provide 75% of the country's electrical needs. Therefore, the People's Republic of China (PRC) must implement CCS to generate low-carbon energy (Li et al. 2015; Korsbakken et al. 2016; Wara 2007; Hu et al. 2017). In the simulation work of Hu et al. (2017), using the Integrated Environmental Control Model (IECM) initially developed by Carnegie Mellon University, the cost of amine-based CCS for new coal-fired power plants in PRC (a nation noted to have contributed roughly 28% of global CO_2 emissions in the year 2013) was studied to vary with major parameters such as plant location, coal price, capacity factor, CO_2 removal efficiency, and fixed charge factor. The techno-economic analysis results for the PCC (with 90% carbon removal efficiency) retrofitted to a 2 × 660 MW supercritical (SC) power plant with a carbon emission capacity of 0.778 kg/kWh are shown in Table 4a. As seen in the table, the integration of CCS lowers the net plant efficiency by 73.6%, resulting in an LCOE increase of 27.5 $/MWh and CCA of 41 $/t$CO_2$ captured. Based on the base case values obtained, various sensitivity analyses were conducted to understand the effects of variations in factors, including plant location, coal quality and price, capacity factor, CO_2 removal efficiency, and fixed charge factor on the LCOE and CCA. At a 95% confidence level

Table 4 TEA analyses of a 1320 MW coal-fired power plant retrofitted with CCS in China (Hu et al. 2017)

Variable	Plant without CCS	Plant with CCS
Gross electrical output (MWg)	1320	1497
Net power output (MWnet)	1238	1238
Net plant efficiency (HHV, %)	42.8	31.5
Coal flow rate (t/hr)	378	514
CO_2 emission rate (kg/kWh)	0.778	0.106
Plant TCR (2013$/kWnet)	635	1160
Plant VOM (2013$/MWh)	27.1	45.9
Plant FOM (2013$/MWh)	1.9	3.5
Plant LCOE (2013$/MWh)	37.6	65.1
Added plant TCR for CCS (2013$/kWnet)		517
TCR of CO_2 capture system alone (2013$/kWnet)		353
Cost of CO_2 avoided (2013$/t$CO_2$)		41.0
Increase in plant LCOE for CCS		
(2013$/MWh)		27.5
(%)		73

HHV = Higher Heating Value.

(CL), these analyses revealed that adding CCS for 90% CO_2 capture could increase the cost of electricity generation by nearly 58%–108% compared to the plant without CCS. At the same CL, the CCA was reported to be within the interval of $35–$67/tCO_2 captured, a cost recognised by the authors to be lower than those obtained in other countries (Table 5.

The quality and price of coal are known to vary by location. The HHV values of the selected coals used in the sensitivity analysis by Hu et al. (2017) range from 19,500 kJ/kg to 28,900 kJ/kg, while the coal prices range from $32 to $100/t at different locations. For a given coal, the extra LCOE due to the added CCS facility and the CCA rises from $23 to $34/MWh and from $34 to $50/t when the coal price fluctuates from $1.3 to $4.9/GJ, respectively. Regarding the impacts of variations in plant capacity factor (CF) and fixed charge factor (FCF), the parametric analysis shows that the CCA rises from $40/t to $48/t when the CF value reduces from 85 to 55% at an FCF value of 10.2%. At a constant CF value of 85%, both the added LCOE for CCS and CCA increase substantially by 26% when the value of FCF changes from 6 to 15%. When the effects of variations in the CCS capital cost adjustment factor (CAF), a function of the type of subsystem adopted, were evaluated, the findings indicated that the CCA rises from $37 to $50/t$CO_2$ when the CAF varies from 0.2 to 0.8. The CO_2 capture efficiency was also directly proportional to the added LCOE for CCS but inversely proportional to the CCA. The summarised TEA results of some other works for CCS integrated into Chinese coal-fired power plants are shown in Table 5. As observed from the table, the cost of CO_2 avoided is generally lower in China as compared to other countries. This observation may be due to lower CAPEX, lower OPEX and lower coal costs in the PRC.

4.2 CCS Retrofitted to a Gas-Fired Power Plant

The techno-economic performance of an amine-based carbon capture and storage (CCS) unit connected to a natural gas combined cycle power plant (NGCC) and a supercritical pulverised coal (SC PC) power plant in 13 different nations was compared in Ferrari et al.'s work in 2019. Their task was to evaluate the differences in TEA performances between the NGCC and SC PC plants retrofitted with a post-combustion CCS facility, depending on their locations. According to their findings, the availability of inexpensive natural gas is the primary factor influencing the variances in the LCOE of NGCC plants with CCS facilities, which vary from 48 to 93 €/MWh among the 13 countries. In the case of the SC PC plant with CCS unity, the LCOE varies between 68 and 106 €/MWh, with variations mainly influenced by coal price, labour costs, and material costs. Generally, installing a CCS facility resulted in added LCOE within the range of 40–70% for NGCC plants and 80–100% for SC PC plants. The CO_2 avoidance cost was reported within the range of 62–95 €/tCO_2 for NGCC plants with CCS facilities and 48–74 €/tCO_2 for SC PC plants with CCS units. Although the Netherlands was chosen as the reference nation, location-specific

Table 5 TEA results for CCS retrofitted to coal-fired power plants in China (summarised by Hu et al. 2017)

Parameter	Li et al. 2011	Wu et al. 2013	Dave et al. 2011	Jin et al. 2015
Reference Plant Without CCS				
Gross power output, MWg	–	600	600	600
Net power output, MW	1000	–	570	570
Net plant efficiency, HHV, %	41	38	41	41
Cost year	2010$	2010$	2010$	2014$
Coal price ($/GJ), HHV	4.9	4.6	3.6	3.3
Plant TCR, $/kW net	664	1106	688	794
Plant FOM, $/MWh	5.92	4.62	–	–
Plant LCOE, $/MWh	–	–	42.8	52
Plant With Retrofitted CCS				
Carbon removal efficiency, %	84	90	90	90
Net power output, MW	–	–	412	389
Net plant efficiency, HHV, %	–	27	30	28
TCR for CCS, $/kWnet	398–445	674	587	636
Plant FOM, $/MWh	3.1–3.5	11.7	–	–
LCOE for CCS, $/MWh	–	–	21.3	46.3
CCA, $/t$CO_2$ captured	39–47	–	31	52

In the reference article (Hu et al. 2017), Jin et al. (2015) were cited as ADB (2015).

Table 6 Cost factors compared to reference plant in The Netherlands (Ferrari et al. 2019)

Location	Material factors	Labour factors (productivity)	Labour factors (costs)	Contingency (%)
Netherland	1.00	1.00	1.00	10
Eastern Europe	0.92	1.28	0.40	10
USA (mid-west)	0.94	0.95	0.92	10
Canada	1.07	1.12	0.96	10
South Africa	1.03	2.24	0.70	10
Australia	1.00	1.23	1.38	10
India	0.93	2.42	0.26	15
China	0.77	2.29	0.16	15
Japan	0.91	0.98	0.68	10
South-East Asia	0.92	1.78	0.24	15
South America	0.97	1.54	0.28	20
Middle East	0.91	1.84	0.24	20

cost factors for the different countries were used in the techno-economic evaluations. Details on these factors can be seen in Table 6.

The location-specific analysis results for the NGCC and SC PC plants with and without CCS facilities are shown in Figs. 1 and 2. Based on the results, the following observations could be made:

- In the European market where natural gas is imported, without CCS technology, NGCC is only a little more competitive than SC PC plants, with roughly 10% less LCOE. Whereas the total investment cost may be lower, this lower cost is partially offset by the higher natural gas price.
- However, integrating a CCS facility leads to an 80% LCOE increase for SC PC against a much lower increment of 45% for NGCC plants. The higher specific carbon content influences the higher LCOE penalty for coal than natural gas, ultimately leading to higher investment costs and CO_2 T&S costs.
- In the cases of SC PC plants, the highest LCOEs are associated with locations in Australia, South Africa, and Canada. This is due to the high investment and coal costs compared to the reference case in the Netherlands. The

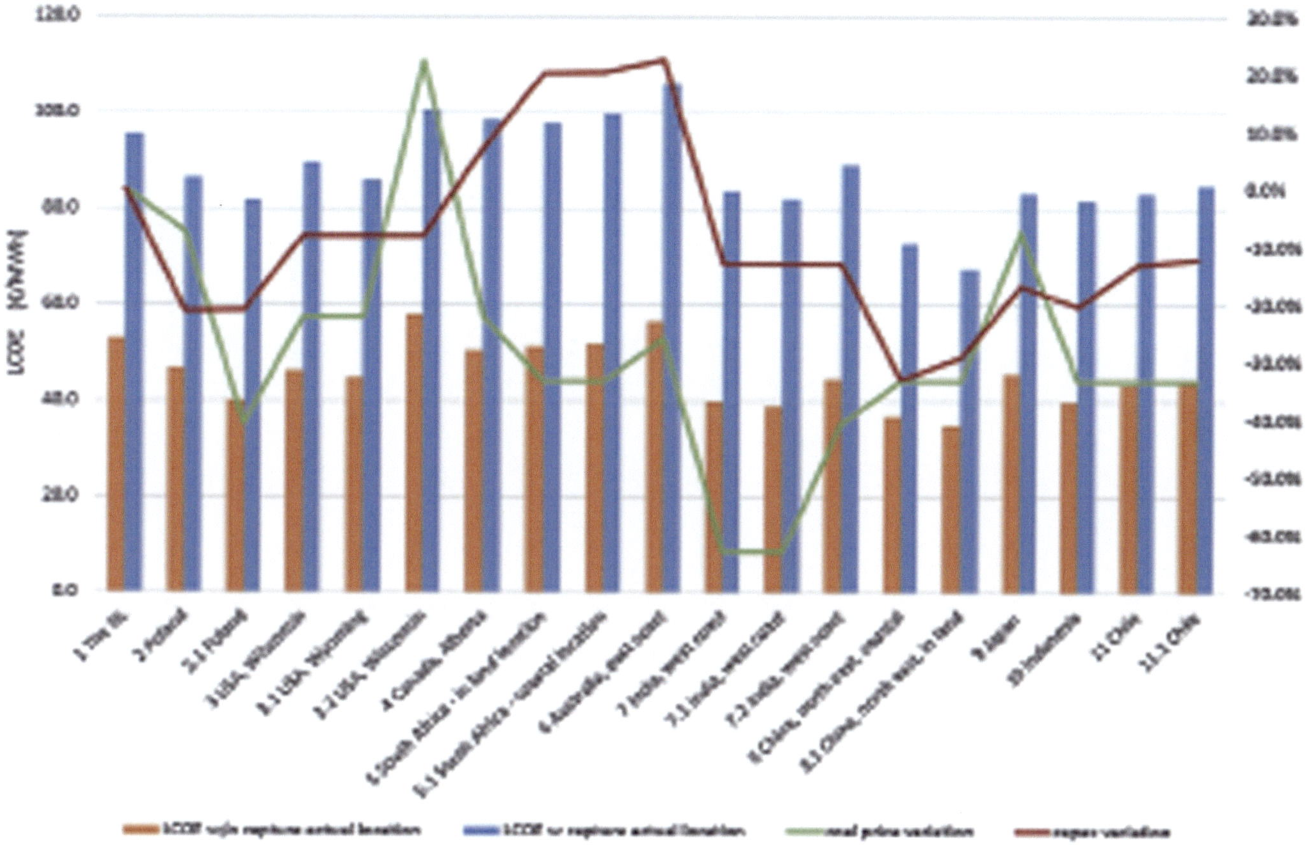

Fig. 1 LCOE of SC PC cases vs total plant cost and coal price (Ferrari et al. 2019)

lowest LCOEs are in China, due to the lower investment costs, and in Poland and India, where cheap local coal sources are readily available.

- Without a CCS facility, the LCOE of SC PC plants ranges from 35 to 58 €/MWh and from 67 to 106 €/MWh for the plants with a CCS unit.
- The impact of fuel price on LCOE is especially pronounced in NGCC cases. Without a CCS facility, the LCOE ranges from 28 to 66 €/MWh and from 48 to 93 €/MWh for the cases with capture units.

The impacts of plant location and plant type on the CCA are illustrated in Figs. 3 and 4. Considering this parameter, SC PC cases appear more convenient than NGCC cases, with a 60–95 €/t CAC against a 50–75 €/t for the former.

- CAC in SC PC cases typically depends on the capital cost, with a minimal impact associated with the specific fuel carbon content and, ultimately, the CO_2 capture level and the penalty for capture. Whereas the lowest CAC (around 50 €/t) is found at locations in China, the highest costs (70–75 €/t) are related to South Africa and Canada.
- On the other hand, the CAC in NGCC plants is influenced by capital costs and natural gas costs. The lowest CAC is found in the Chinese regions, characterised by the lowest fuel costs, and the highest is associated with the locations in South Africa, characterised by high natural gas costs.

According to the authors, technical site variations and economic aspects are the primary factors influencing plant performance and costs. While fuel analysis and ambient conditions are the primary factors influencing technical differences between the sites, fuel prices, labour costs and productivity, and local material costs are the key economic factors (Ferrari et al. 2019).

5 TEA of CCS Systems Integrated to Other Chemical and Industrial Plants

While CCS literature contains several works on the TEA for capture plants designed for processing emissions from power plants, there are only a few such studies for CCS systems integrated into other industrial plants. This section covers the analysis of CCS units attached to steel mills, aluminium factories, cement plants, and pulp mills.

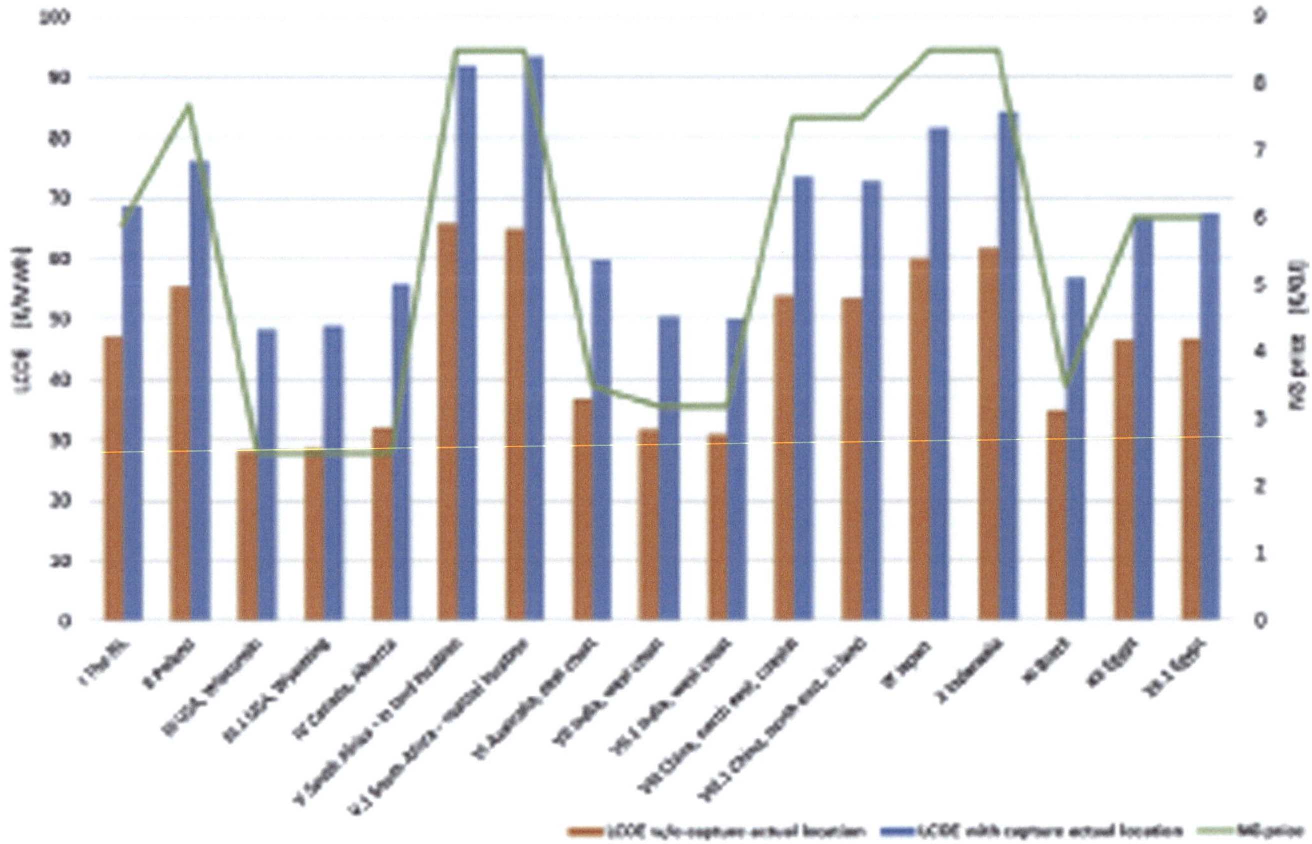

Fig. 2 LCOE of NGCC cases vs natural gas price (Ferrari et al. 2019)

5.1 CCS Integrated to a Steel Mill

In a simulation-based study by Arasto et al. (2013), which analysed the technical concept of a post-combustion CO_2 capture at an integrated steel mill in Finland (an emission source noted to release the most significant amount of CO_2 in the country, approximately 4 Mt/year), the authors investigated the technical implications of varying CO_2 capture rates, different capture solvents for the separation process and different heat supply possibilities for solvent regeneration. The focus of the project was to understand how the above factors impact the steel mill's energy balance and greenhouse gas emissions. Their findings show that the chemical absorption process can capture 50–70% of CO_2 (accounting for 2–3 Mt CO_2/annum). However, the authors noted that capturing a large amount of the emissions is technically and economically less feasible because of the large number of small stacks at the vast production site.

Using a standard 30wt% MEA (widely considered the baseline solvent) as the absorbent in the post-combustion CC system, the authors reported a solvent regeneration duty of 3.4 MJ/kg captured CO_2. Using amino acid salt solution as the capture solvent resulted in a lesser solvent regeneration energy consumption of 2.7 MJ/kg captured CO_2. The solvent regeneration energy requirement for the system using an imaginary solvent (simply referred to as low-T because of its potential to be regenerated at a significantly lower temperature than conventional MEA) was reported as 3.0 MJ/kg captured CO_2.

Further, the authors also noted that electricity generation at the reference power plant supplying electricity to the steel mill decreased by 40% from 1200 GWh/yr to 730 GWh/yr when low-pressure steam, which could have been utilised in the steam turbine, is utilised for solvent regeneration in the carbon capture plant.

A pitfall of this article, as is the case in many other literature publications on TEA for CCS, was that evaluations on the power needed in the capture process for pumps and blowers for flue gas and solvent recirculation, as well as the cost estimate for the compression and T&S of the captured CO_2, was not delineated. However, a rough estimate of 0.41 MJ/kg captured CO_2 was reported as the total power consumption accounting for the power usage of pumps, compressors and other auxiliaries, with CO_2 compression singled out as the most power-intensive process. Regarding the storage of the captured CO_2, the authors recognised that the capacity for permanent geological storage of CO_2 is not available in Finland, requiring the compressed CO_2 to be transported to a

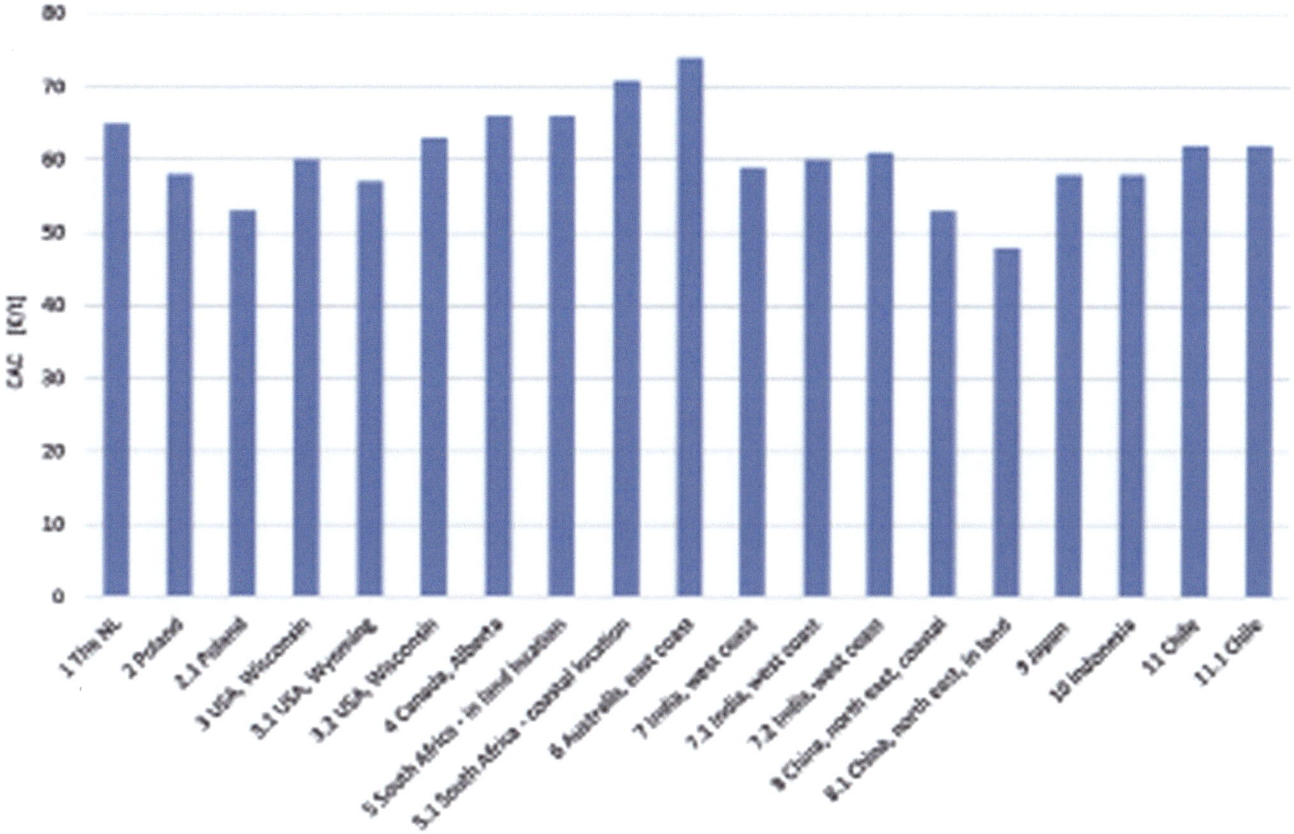

Fig. 3 Cost of CO_2 avoided for SC PC cases (Ferrari et al. 2019)

storage facility outside the country's borders. To achieve this, ship transportation was identified as the most economical. This requires the captured CO_2 to be pressurised to 6.5 bar and −52 °C, conditions considered suitable for safe transport. Techno-economic analysis on the T&S of the CCS technology was completely lacking in this report.

In a sequel to the above publication (Tsupari et al. 2013), the authors delved more into the economic evaluations of the solvent-based post-combustion CC plant integrated into the same steel mill, using a Microsoft Excel-based system model called CC-Skynet. At COE of 100 €/MWh and the European Union Emission Trading Scheme's (abbreviated as EU-ETS or EUA) increased steel mill price of 80 €/t, their economic analysis results revealed that the break-even price (BEP) of CCS is approximately 72 €/t CO_2 for the MEA-based capture system operating at 50–70% capture rate (2–3 Mt CO_2/annum). At the same COE of 100 €/MWh and EUA's increased steel mill cost of 80 €/t, the BEP of CO_2 emission allowance for an amino acid salt solution is nearly 64 €/t CO_2. Different values (approximate) of the BEP at varying COE for baseline MEA and amino acid salt solution are presented in Table 7. According to the authors, the BEP values for the Low-T solvent are close to those for the amino acid salt solution (AASS).

In the same paper, the authors estimated the CO_2 avoided cost for an MEA-based capture plant to range from 94 to 186 €/t at COE of 80 €/MWh and from 95 to 212 €/t at COE of 100 €/MWh. To complete this analysis, the electricity supply to the steel mill was assumed to be generated by a coal-fired power plant, and the variations in the CCA were recognised to be heavily dependent on the characteristics of the capture facility and the operational environment. When the electricity supply to the steel mill was assumed to be generated from a natural gas-based power plant, the range of CO_2 avoided cost decreased to 84–119 €/t at COE of 80 €/MWh. However, the authors did not report the estimates at COE of 100 €/MWh for the latter scenario.

5.2 CCS Coupled to an Aluminium Factory

In the simulation-based study published by Jilvero et al. (2014a, b), NH_3 and MEA were assessed for PCC at one section of an aluminium manufacturing plant with an annual carbon emission capacity of 0.42 Mt in Norway. The study focused on the technical performance and economic feasibility of the retrofitted capture plant as a function of flue gas CO_2 concentration (vol %). At a carbon capture efficiency

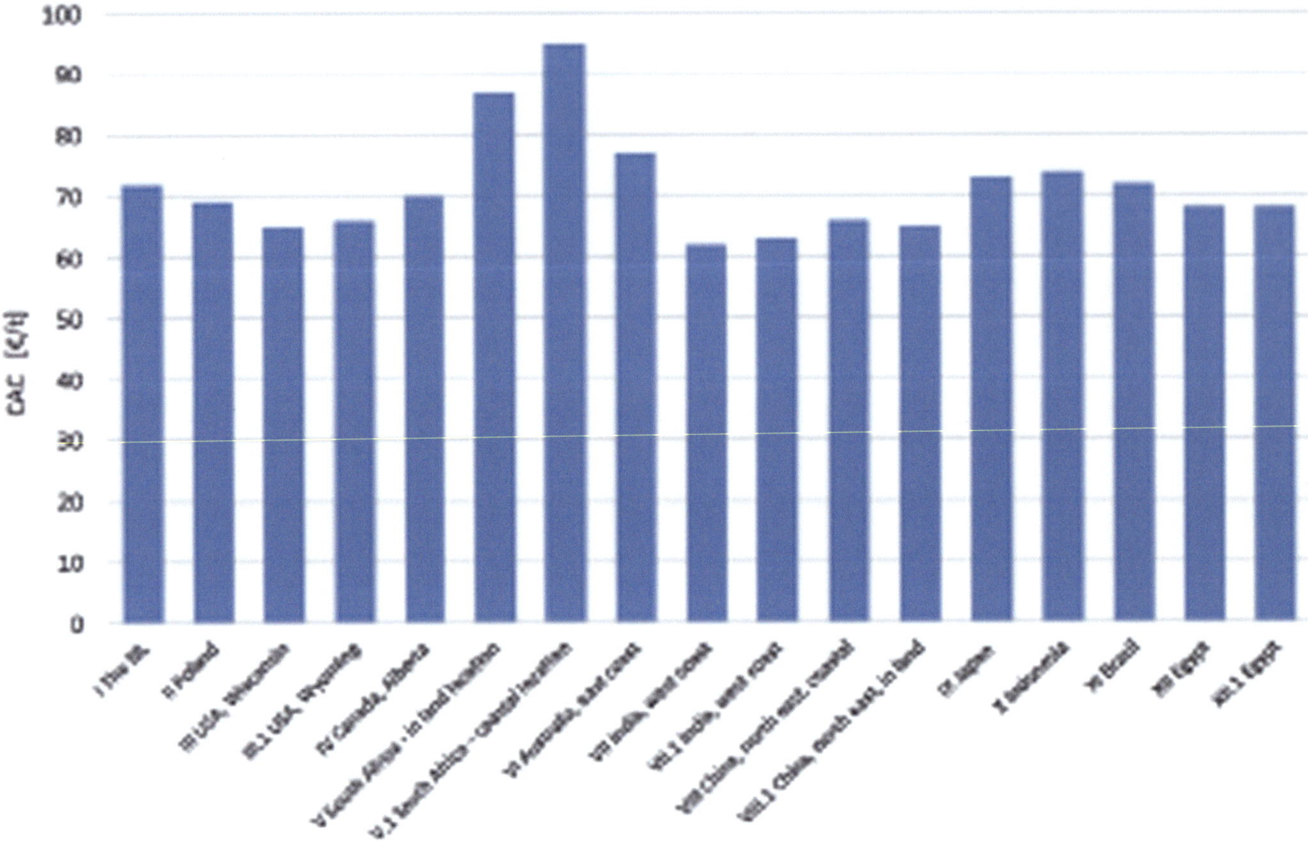

Fig. 4 Cost of avoided CO_2 for NGCC cases (Ferrari et al. 2019)

Table 7 BEP as a function of COE based on Tsupari et al. (2013)

		coe, €/MWh										
		50	60	70	80	90	100	110	120	130	140	150
BEP,	MEA	55.5	59	62.5	65.2	69	72	75.2	79	82	85.5	89
	AASS	49.5	52	55	57.5	61	64	66.5	70	72.5	75	78

of 85%, the specific heat requirements for solvent regeneration and external heat consumption for both processes were technically estimated as functions of CO_2 concentration, as shown on the clustered column charts in Figs. 5 and 6, respectively. Although both absorbents showed fairly equal specific heat requirements for solvent regeneration with respect to changes in CO_2 concentration in the flue gas, the authors noted that a decreased CO_2 concentration takes a major toll on the heat requirement of the stripper column for the NH_3-based process due to the sharp increase in energy duty for the ammonia abatement cycle for 4% CO_2 concentration and below. When the two processes were compared concerning the specific capture cost, the results revealed that the NH_3-based process is the most cost-efficient at high CO_2 concentrations (7–10%), and the MEA-based process is the most cost favourable option at low CO_2 concentrations (3–4%). The CCCs for both processes were reported as 74–82 €/tCO_2 captured for the NH_3-based system and 93–97 €/tCO_2 for the MEA-based plant. A heat integration analysis shows that up to 50% of the heat requirement could be covered with excess heat from the aluminium manufacturing plant, which could lower the specific carbon capture cost by 10–15 €/tCO_2 captured.

5.3 CCS Coupled to a Cement Factory

Although few works have considered the techno-economic performances of CCS technologies deployed to decarbonise cement manufacturing plants, few works exist in the literature. 'A techno-economic case study of CO_2 capture, transport and storage chain from a cement plant in Norway', an article

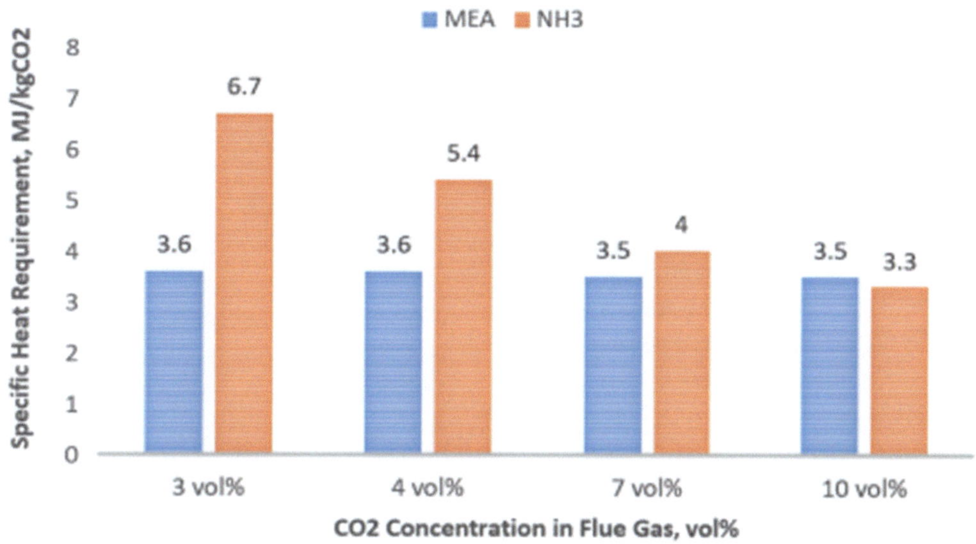

Fig. 5 Specific heat duty as a function of CO_2 concentration (basec on Jilvero et al. 2014a, b)

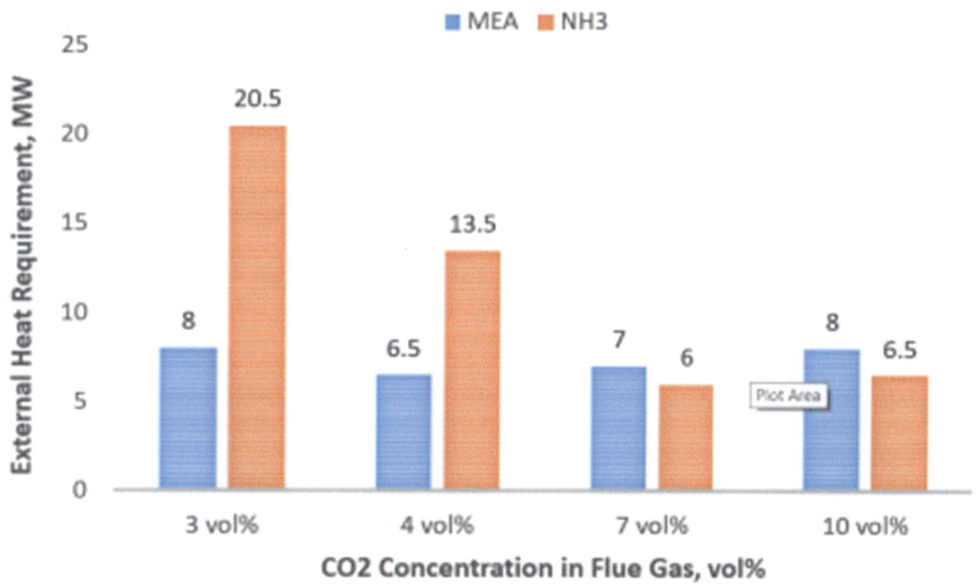

Fig. 6 External heat requirement as a function of CO_2 concentration (based on Jilvero et al. 2014a, b)

published by Jakobsen et al. (2017), is one such research work which used an in-house techno-economic assessment tool to evaluate the technical and economic performance of a CCS integrated to the Norwegian Norcem Brevik cement plant. The carbon emission capacity of this plant is approximately 925 kt/y of CO_2. The software used for the TEA was simply referred to as iCCS, which SINTEF Energy Research developed within the BIGCCS Research Center (operational from 2008 to 2016), the predecessor to the current Norwegian CCS (NCCS) Research Center for Environmental-Friendly Energy Research (operational from 2016 to 2024) (Aarlien 2017). A matrix of nine (9) CCS alternative chains was analysed to quantify the cost reduction potentials across various CCS value chains. A summary of the chain design options used is shown in Table 8.

Chain A is considered the baseline design (uses 30 wt% MEA), with a capture efficiency of 42% and solvent regeneration energy usage of approximately 3.2 GJ/tCO_2. The same capture efficiency is considered for chains B–H. However, the solvent regeneration energy requirement was assumed to be 2.1 GJ/tCO_2 in chain B, which uses an advanced solvent for the carbon separation process. The only chain I was operated at a full-scale carbon removal capacity of 85%. The overall results of the cost of CO_2 avoided for the nine different CCS chain alternatives are summarised in Fig. 7. The results indicate that the full-chain cost of the base case is estimated to be around 120 €/tCO_2 avoided, whereas the alternative chains cost between 100 and 150 €/tCO_2 avoided. In the base case, CO_2 capture and conditioning represent roughly half of the CAC, whilst transport and storage account for 25% and 20%

Table 8 Summary of chain design options (Jakobsen et al. 2017)

Chain	Capture	Transport	Storage	Storage location
A	MEA-based capture	Shipping	Saline aquifer (SA)	L1
B	Advanced solvent-based capture	Shipping	Saline aquifer	L1
C	Membrance-based capture with Polaris membrance	Shipping	Saline aquifer	L1
D	Membrance-based capture with FSC membrance	Shipping	Saline aquifer	L1
E	MEA-based capture	Pipeline	Saline aquifer	L1
F	MEA-based capture	Shared pipeline	Shared saline aquifer	L1
G	MEA-based capture	Shipping	Shared EOR storage over 25 years with nearby shared SA for excess CO_2	L2
H	MEA-based capture	Shipping	Shared EOR storage the first 10 years and shared SA for excess CO_2 and after 10 years	L2
I	MEA-based capture at full scale	Shipping	Saline aquifer	L1

- FSC = Fixed Site Carrier
- EOR = Enhanced Oil Recovery
- L1 = The Gassum formation covers large parts of the Skagerrak and south towards Denmark. Identified as the most attractive storage option in the area and the most viable option for storage of CO_2 from point sources in eastern Norway. This storage site lies about 300 km from the Norcem cement plant and is believed to be only suitable for storage in an aquifer.
- L2 = Johansen formation in the northern part of the North Sea. It is a water-filled reservoir about 500–600 m deep, beneath which a porous reservoir produces oil and gas from the Troll field. It is considered the most cost-efficient CO_2 storage site between the Utsira Formation (south of Sleipner) and Johansen. This option is roughly 450 miles from the Norcem plant and is suitable for only CO_2 storage in an aquifer or combined with an EOR project.

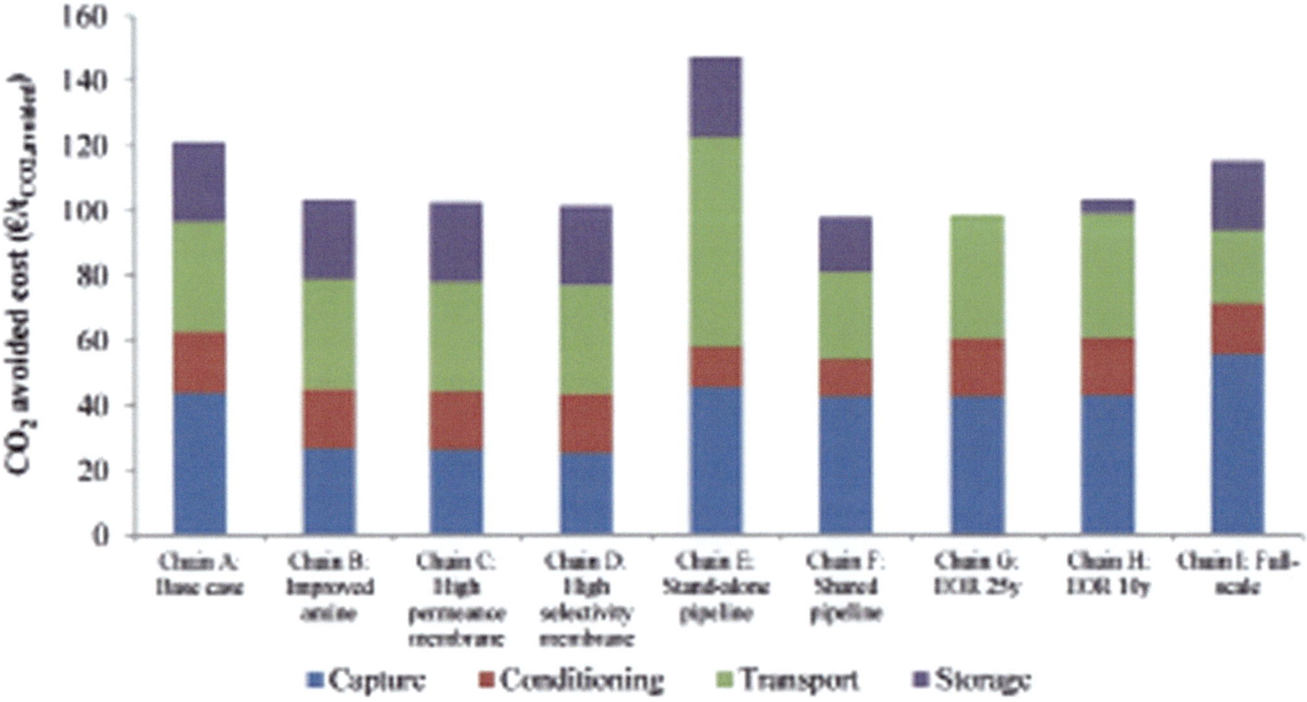

Fig. 7 Cost of CO_2 avoided for the nine CCS alternative chains (Jakobsen et al. 2017)

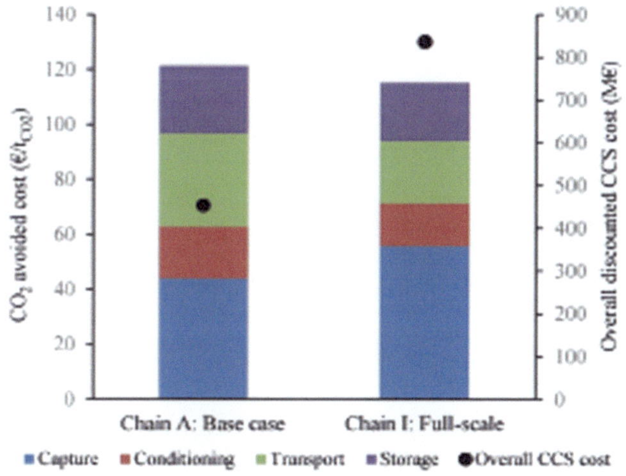

Fig. 8 The cost of CO_2 is avoided for the base case and the full-scale case (Jakobsen et al. 2017)

of the cost. These proportions are specific and depend on the characteristics of the chain considered.

Figure 8 compares the cost performance of the base case scenario with partial operation scale to the full-scale operation. The results indicate that while full-scale CCS is often aimed to demonstrate and implement CCS technologies due to its assumed potential economies of scale, it might not always be the most suitable target for all implementation cases. In the case under review, a full-scale demonstration might reduce the CAC by only 6% compared to the partial scale. Even if the whole chain benefits from the economies of scale related to the higher CCS capacity, the CO_2 capture cost rises significantly in the full-scale case. This is so because, in the base case scenario, a large portion of the steam the capture unit requires can be generated cheaply from the cement plant's waste heat. However, an external energy source is required for the same purpose in the full-scale scenario, increasing the CAC due to the additional steam required from an NGCC or coal-fired power plant. This offsets the economies of scale in the case of a full-scale facility.

The cost performance of the three different transport scenarios studied in this work is compared in Fig. 9. Aside from its advantage of being a continuous process with opportunities for automation, offshore pipeline transport is often acknowledged as the most convenient due to its cost-effectiveness for transporting large volumes of CO_2. Nonetheless, a drawback of pipeline transport is that it entails large initial capital investments. Whereas CO_2 shipping is more flexible and less capital-intensive, on the other hand, it is not cost-effective for short distances and large volumes.

A comparison of the stand-alone pipeline transport with the shipping base case shows a rise of 45% in the cost of CO_2 conditioning and transportation. Even if the conditioning costs are lower in the case of pipeline transport (Roussanaly et al. 2013), pipeline CO_2 transportation is a highly cost-inefficient way of transporting small volumes over long distances (Knoope et al. 2013).

However, when evaluating a common pipeline infrastructure shared with two other actors, the costs of transport and conditioning allocated to the cement plant experience a 50% reduction compared to the stand-alone case. Consequently, conditioning and transport costs decrease by 30% over the base case scenario, assuming shipping is the primary mode of transportation.

The possible cost benefits of CO_2 storage related to EOR were compared to the base case scenario. The assessment outcomes indicated that significant cost reductions could

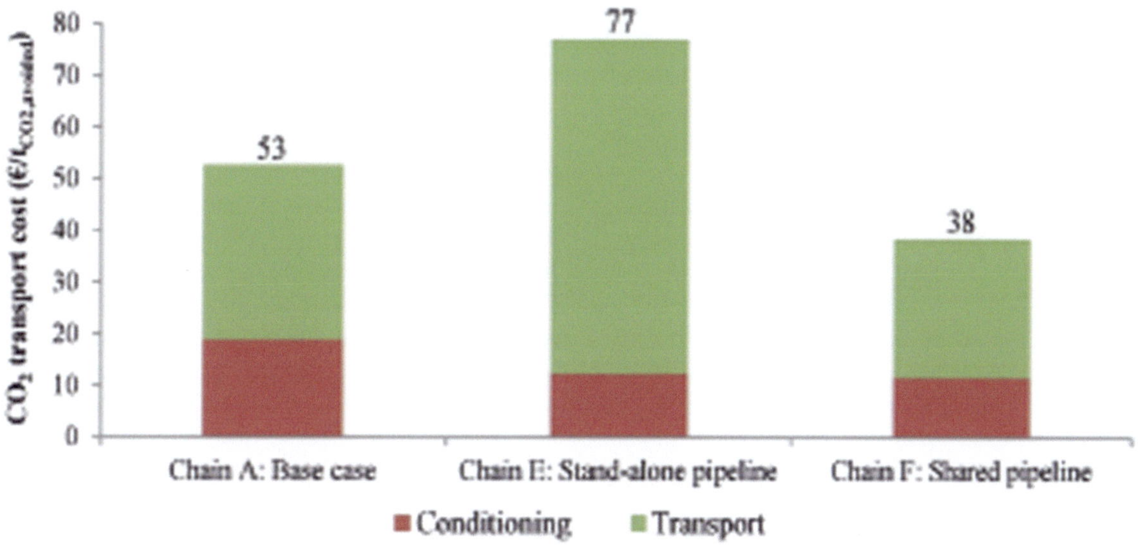

Fig. 9 CO_2 conditioning and transport cost of the base case with shipping and the two pipeline transport alternative cases (Jakobsen et al. 2017)

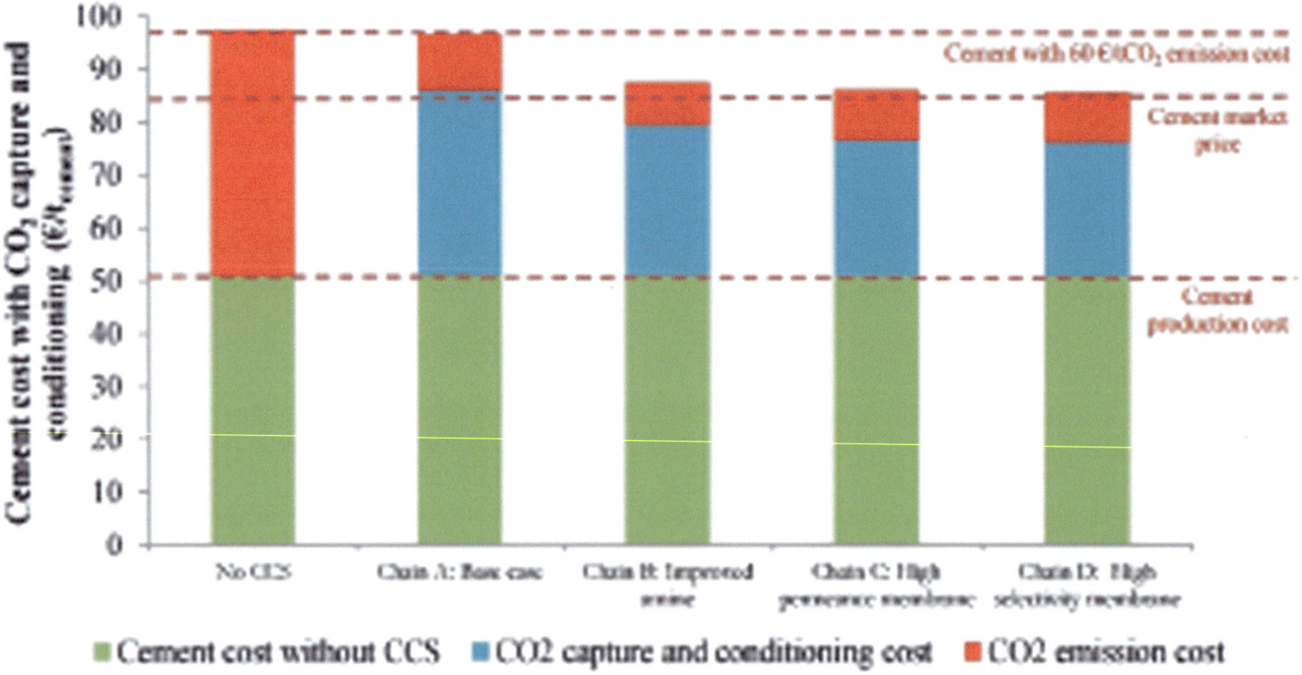

Fig. 10 Cement cost with CO_2 capture and conditioning for the different capture alternatives at a CO_2 emission cost of 60 €/tCO_2 (Jakobsen et al. 2017)

be achieved if CCS projects are coupled with CO_2 EOR storage under the oil prices considered. Truly, the costs of transporting and storing CO_2 could be reduced by 35% and 28%, respectively, in the cases of 25 and 10 years of EOR storage. However, it is worth noting that the viability of CO_2 EOR storage highly depends on the oil value considered (Roussanaly and Grimstad 2014).

Finally, the impact of integrating a CCS facility on the cost of cement was assessed. Figure 10 illustrates that incorporating the expenses related to CO_2 capture and conditioning into the production costs of cement results in a notable increase of 70%. This rise, not factoring in transportation and storage expenses, surpasses producers' profit margin, rendering the adoption of CCS in cement production economically unfeasible at present market prices unless suitable compensation measures are implemented.

Therefore, should the implementation of CCS technology at the Norcem cement plant be pursued, it would necessitate financial backing from the public sector to offset the additional costs associated with integrating CO_2 capture and conditioning processes at the facility. The extent of financial support required would be contingent upon the specific capture technology employed, with estimates ranging from 40 €/tCO_2 avoided using advanced amine and membrane-based capture methods to 60 €/tCO_2 avoided utilising the conventional MEA process.

5.4 TEA for Other CCS Facilities

Other studies on CCS technology attached to various emission sources, which have not been discussed previously, are presented in Table 8. The summary of the results from these studies is covered in this section.

Based on the results from various cost-optimal analyses completed by Alhajaj et al. (2016) for a CCS facility attached to an NGCC power plant, the authors posited that unless there is a sufficiently high carbon price, an amount higher than $60/t$CO_2$, which would be enough to absorb the levelised cost of CO_2 capture and compression (LCCC), it would be more advantageous to vent the CO_2 rather than investing in CCS. The minimum LCCC recorded in this study was $63/t$CO_2$.

In a subsequent investigation by Zhang et al. (2012), the TEA was executed to evaluate three distinct coal types: anthracite, bituminous coal, and lignite. The findings indicate a reduction in net power outputs of the retrofitted power plant, decreasing from 302.62 to 249.90 MWe for anthracite, 254.97 MWe for bituminous coal, and 249.45 MWe for lignite, respectively. Overall, there was an approximate 7% decline in net efficiencies across the board. Concurrently, the total annual cost of the power plant equipped with a retrofitted CCS unit exhibited an increase from 80.6 M$/year to 88.8 M$/year for anthracite, 90.8 M$/year for bituminous coal, and 87.5 M$/year for lignite. Additionally, the levelised cost of electricity (LCOE) was projected to escalate from 37 $/MWh to 48.4 $/MWh for anthracite, 50.5 $/MWh for bituminous coal, and 48.7 $/MWh for lignite. Furthermore, the carbon capture and abatement (CCA) costs for anthracite, bituminous coal, and lignite were anticipated to be 15.2 $/t$CO_2$, 19.6 $/t$CO_2$, and 15.0 $/t$CO_2$, respectively.

In another study by Liang and Li (2012), which focussed on a CCS unit attached to a cement plant, the researchers operated under the assumption that the additional energy and heat required for CO_2 capture and compression would be supplied by a newly established 200 MW combined heat and power (CHP) unit. With a remaining operational lifespan of 25 years and a discount rate of 14%, the projected CCA through retrofitting the CCS unit to the cement plant was determined to be $70 per metric ton of CO_2 (excluding costs associated with CO_2 transport and storage). Utilising a stochastic cash flow analysis coupled with a real option model and Monte Carlo simulation, the study determined the option's value to retrofit to be $1.2 million, with a corresponding probability of economic viability of 7.3%. The authors underscored the sensitivity of their estimation to assumptions within the carbon pricing model, noting a baseline carbon price of $12.00 per metric ton of CO_2 in 2012, escalating at a mean growth rate of 8%. Should a 10% mean growth rate in carbon pricing be adopted, the option value and probability could escalate to $20 million and 67%, respectively. Considering these findings, the authors concluded that retrofitting cement plants with a PCC facility is comparatively less appealing than retrofitting such a unit to power plants.

One of the novel articles reviewed was by Nwaoha and Tontiwachwuthikul (2019). This study investigated the techno-economic performance of a solvent-based PCC plant employing MEA and AMP, integrated with a pulp mill producing 365,000 air dry tons (ADt) of pulp annually in British Columbia, Canada. Various configurations of the PCC facility were assessed, including the Conventional process (CONV), advanced rich amine 4-split (ARA4S), and advanced rich amine 3-split with desorber inter-heating (ARA3S-DI). Comparative analysis indicated that the capture cost of a 5 kmol/m^3 MEA system was the lowest for the CONV configuration. Conversely, for a blended solution containing 2 kmol/m^3 AMP and 3 kmol/^{3} MEA, the lowest cost was observed for ARA3S-DI. Additionally, the capture costs associated with the MEA solution using CONV were 8.7% higher compared to the AMP-MEA blend with ARA3S-DI. Sensitivity analysis revealed that the introduction of a carbon tax ($40/t$CO_2$), CO_2 sales price ($40/tCO_2), and an advanced PCC configuration led to a marginal increase (2.6% for MEA and 2.1% for the AMP-MEA blend) in the price of northern bleached softwood kraft (NBSK) pulp. These findings suggest that, with an improved solvent system and process configuration, the price of NBSK pulp may remain unaffected under the prevailing CO_2 tax and pricing regime.

Lassagne et al. (2016) examined the techno-economic viability of integrating a PCC unit with an aluminium production facility employing PZ and AMP as capture solvents. The study explored varying concentrations of PZ (ranging from 0.02 to 0.10 g/g) and AMP (ranging from 0.30 to 0.38 g/g) to maintain a constant total amine concentration (TAC) of 0.4 g/g in the solution. In addition to assessing the economic implications of CO_2 capture, the research investigated the feasibility and expenses associated with integrating the capture process with a waste heat recovery system. Results indicated that a blend of 0.08 g/g PZ and 0.32 g/g AMP represented an optimal configuration, reducing capital costs by approximately 25% and operational costs by around 29% compared to MEA solutions. The estimated capture cost stood at 65.10 $/t of CO_2 avoided (excluding carbon tax). Further analysis demonstrated that incorporating thermal integration of the capture plant within the primary aluminium smelter could potentially lower the cost to 57.57 $/t of CO_2 avoided, or equivalently, 69.56 $/t of aluminium (Al) produced (excluding carbon tax).

In their analysis of a PCC process aimed at decarbonising an NGCC power plant, Osagie et al. (2018) conducted a comparative evaluation between AMP and MEA. Their thermodynamic examination revealed a notable advantage of the AMP-based process, with a 25.6% reduction in reboiler duty compared to the conventional MEA-based approach. This discrepancy stemmed primarily from the ability of the AMP solvent to undergo regeneration at elevated temperature (140 °C) and pressure (3.5 bar), contrasting with MEA's lower regeneration conditions (120 °C and 1.8 bar). Moreover, integrating the AMP-based process with the NGCC power plant incurred a lower efficiency penalty of 7.1% points compared to 9.1% points associated with the MEA-based process retrofitting. Despite its thermodynamic superiority, economic viability favoured the MEA-based process for make-up rates exceeding 0.03%. Consequently, employing AMP as a solvent in the PCC plant may not present the most economically feasible option. Nevertheless, AMP demonstrates notable efficacy in mitigating the efficiency penalties inherent in the capture process.

Nwaoha et al. (2018) conducted the TEA of CO_2 capture from a 1.2 million MTPA cement plant using a blend of AMP, PZ, and MEA (19 wt% AMP-10 wt% PZ-12 wt% MEA). The impact of the amine type, energy penalty, and CO_2 capture efficiency (50% to 90%) on the capture cost were investigated. In addition, sensitivity analysis on the impact of CO_2 capture plant, carbon tax (20–40 $/t$CO_2$), CO_2 sales price (10–40 $/t$CO_2$), energy penalty and CO_2 capture efficiency on the cement price was also investigated. The outcomes indicated that at 90% CO_2 capture efficiency, the capture costs of AMP-PZ-MEA blend, which is estimated at 77.34 $/t$CO_2$ or 44.94 $/t cement, is lower than that of MEA only, projected at 93.23 $/t$CO_2$ or 54.17 $/t cement. Other analysis results also revealed that the total equipment cost and capital expenditure (CAPEX) of the MEA-only system (29.76 $M and 147.12 $M) is greater than that of the AMP-PZ-MEA blend (23.39 $M and 127.59 $M).

Lib et al. (2016) worked on the TEA of an aqueous NH_3-based PCC process integrated into a 650-MW coal-fired power station. Aside from comparing the baseline scenario of the NH3 process to the benchmark MEA process, the authors also implemented process modifications such as a two-stage absorption, rich-split configuration and inter-heating stripping to further advance the NH_3–based capture process. Based on their results, the cost of CO_2 avoided for the base case scenario (NH_3–based capture process) was 67.3 $/t$CO_2$ as against the MEA process cost of 86.4 $/t$CO_2$. The combined modified process reduced energy consumption by 31.7 MW/h (20.2% reduction) and capital costs by 55.4 $M (6.7% reduction). As a result, the CAC decreased to 53.2 $/t$CO_2$, a savings of 14.1 $/t$CO_2$ and 21.9 $/t$CO_2$ compared to the NH_3 baseline and advanced MEA process.

In the publication by Anderson et al. (2014), the authors investigated the economics of a large-scale potassium carbonate (K_2CO_3)–based PCC unit retrofitted to a 550 MWe black coal-fired power plant in Australia. The study used UNO MK3, a K_2CO_3 process developed and patented by the Cooperative Research Centre for Greenhouse Gas Technologies (CO2CRC). Their findings predicted the LCOE to be 73 $/MWh, representing an increase of 24% in the cost of electricity due to the installation of a PCC unit, and the cost of carbon capture as 21 $/t CO_2 avoided. It is worth noting that the authors mentioned that cost estimates exclude CO_2 T&S costs.

6 Conclusions

This paper reviewed the state-of-the-art publications on the technical performance and cost estimates for chemical absorption-based carbon capture technologies. The performance of the CCS facility retrofitted to various emission sources was evaluated. The findings show that the techno-economic analysis for CCS technologies varies with major parameters such as plant location, coal price, emission source, CO_2 concentration in the flue gas, storage location and type, type of transport facility, solvent type, capacity factor, CO_2 removal efficiency, and fixed charge factor. Whereas the estimates provide general ideas about the cost of CCS technologies, most studies were simulation-based, with works at the pilot and commercial scales rare in the literature. Some estimates could be considered misleading. For instance, some works which did not include the cost estimates for the transport and storage of the captured technology projected costs of CO_2 avoided instead of the cost of CO_2 capture. According to experts, the cost of CO_2 avoided can only be quantified if the costs associated with transporting and storing the captured CO_2 are considered. These experts argued that captured CO_2 is never avoided unless it is safely transported and stored away from the environment. Whereas this argument is true, it is also true that the captured CO_2 is avoided if it is utilised in making other products. In such a case, the cost estimates of CO_2 transport and storage may not necessarily be needed to quantify the cost of CO_2 avoided. However, to limit confusion regarding the meaning of these cost estimates, it is necessary for authors to clearly define the estimates in their work to leave no room or little room for readers to assume what they really mean.

Acknowledgements The first author would like to acknowledge the Research Support Program (RSP) from Curtin University, Australia, which provides funding for this Short-Term Postdoctoral Fellowship. Both authors also thank Curtin University Malaysia and Curtin Malaysia Research Institute (CMRI) for the additional research support.

Author Contributions Foster Kofi Ayittey—writing of original draft, conceptualisation, and editing.

Agus Saptoro—conceptualisation, review, and editing.

Declaration of Interests The authors declare that they have no known competing financial interests or personal relationships that could have appeared to influence the work reported in this paper.

Data Deposition All data underlying this article are incorporated into the article.

Abbreviations

CO_2	Carbon dioxide
CCS	Carbon capture and storage
MEA	Monoethanolamine
AMP	2-Amino-2-methyl-1-propanol
PZ	PiperazinePiperazine
NH_3	Ammonia
K_2CO_3	Potassium carbonate
TEA	Techno-economic analyses
IEA	International Energy Agency
DEA	Diethanolamine
GHG	Greenhouse gas
R&D	Research and development
O&M	Operating and maintenance
T&S	Transport and storage
EPRI	Electric Power Research Institute
IEAGHG	International Energy Agency Greenhouse Gas Program
GCCSI	Global Carbon Capture and Storage Institute
DOE/NETL	US Department of Energy's National Energy Technology Laboratory
TRL	Technology readiness level
NETL	National Energy Technology Laboratory
CCA	Cost of CO_2 avoided
CAC	CO_2 avoidance cost

NPV	Net present value
TCC	Total capital cost
FCF	Fixed charge factor
FOM	Fixed operating and maintenance
VOM	Variable operating and maintenance cost
FC	Fuel cost; CF—capacity factor
LCOE	Levelised cost of electricity
IECM	Integrated Environmental Control Model
TCR	Total capital requirement
PFC	Process facilities cost
TPC	Total plant cost
PCC	Post-combustion capture
CAPEX	Capital expenditures
OPEX	Operating expenditures
PRC	People's Republic of China
CL	Confidence level
HHV	Higher heating value
LHV	Lower heating value
NGCC	Natural gas combined cycle power plant
SC PC	Supercritical pulverised coal
FSC	Fixed site carrier
EOR	Enhanced oil recovery
LCCC	Levelised cost of capture and compression
MTPA	Metric tons per annum

References

Aarlien, R. (2017). *BIGCCS final report.*

Alhajaj, A., Mac Dowell, N. and Shah, N. (2016). A techno-economic analysis of post-combustion CO_2 capture and compression applied to a combined cycle gas turbine: Part II. Identifying the cost-optimal control and design variables. *Int. J. Greenh. Gas Control.* **52:** 331–343.

Anderson, C., Ho, M., Harkin, T., Wiley, D. and Hooper, B. (2014). Large scale economics of a precipitating potassium carbonate CO_2 capture process for black coal power generation. *Greenh. Gases.* **4(1):** 8–19.

Arasto, A., Tsupari, E., Kärki, J., Pisilä, E. and Sorsamäki, L. (2013). Post-combustion capture of CO_2 at an integrated steel mill–Part I: Technical concept analysis. *Int. J. Greenh. Gas Control.* **16:** 271–277.

Burk, C. (2018). Techno-economic modeling for new technology development. *Chem. Eng. Prog.* **114(1):** 43–52.

Clarke, L., Edmonds, J., Krey, V., Richels, R., Rose, S. and Tavoni, M. (2009). International climate policy architectures: Overview of the EMF 22 International Scenarios. *Energy Econ.* **31:** S64-S81.

Clarke, L., Kyle, P., Wise, M., Calvin, K., Edmonds, J., Kim, S., Placet, M. and Smith, S. (2008). CO_2 emissions mitigation and technological advance: An updated analysis of advanced technology scenarios. *PNNL Report Pacific Northwest National Laboratory, Richmond.*

Cooper, C. (2009). A technical basis for carbon dioxide storage. *Energy Proc.* **1(1):** 1727–1733.

Cooper, C. and Pillips, C. (Eds.). (2009). *A technical basis for carbon dioxide storage - UNFCCC.* CO_2 Capture Project. pp 13–15. https://unfccc.int/resource/docs/2011/smsn/ngo/276.pdf.

Dave, N., Do, T., Palfreyman, D., Feron, P.H.M., Xu, S., Gao, S. and Liu, L. (2011). Post-combustion capture of CO_2 from coal-fired power plants in China and Australia: An experience based cost comparison. *Energy Proc.* **4:** 1869–1877.

Deshmukh, M.K.G., Sameeroddin, M., Abdul, D. and Sattar, M.A. (2023). Renewable energy in the 21st century: A review. *Mater. Today Proc.* **80:** 1756–1759

Desideri, U. and Antonelli, M. (2014). A simplified method for the evaluation of the performance of coal fired power plant with carbon capture. *Appl. Therm. Eng.* **64(1–2):** 263–272.

Electric Power Research Institute (EPRI). (1993). *TAGTM, Technical Assessment Guide, Volume 1: Electricity Supply – 1993.* TR-102276-V1R1, Electric Power Research Institute, Palo Alto, CA.

Ferrari, N., Mancuso, L., Burnard, K. and Consonni, F. (2019). Effects of plant location on cost of CO_2 capture. *Int. J. Greenh. Gas Control.* **90:** 102783.

Figueroa, J.D., Fout, T., Plasynski, S., McIlvried, H. and Srivastava, R.D. (2008). Advances in CO_2 capture technology—the US Department of Energy's Carbon Sequestration Program. *Int. J. Greenh. Gas Control.* **2(1):** 9–20.

Guandalini, G., Romano, M.C., Ho, M., Wiley, D., Rubin, E.S. and Abanades, J.C. (2019). A sequential approach for the economic evaluation of new CO_2 capture technologies for power plants. *Int. J. Greenh. Gas Control.* **84:** 219–231.

Hu, B. and Zhai, H. (2017). The cost of carbon capture and storage for coal-fired power plants in China. *Int. J. Greenh. Gas Control.* **65:** 23–31.

IEA, I. (2011). *World Energy Outlook. 2011.* International Energy Agency, 666.

Jakobsen, J., Roussanaly, S. and Anantharaman, R. (2017). A techno-economic case study of CO_2 capture, transport and storage chain from a cement plant in Norway. *J. Clean. Prod.* **144:** 523–539.

Techno-economic analysis of carbon capture at an aluminum production plant–comparison of post-combustion capture using MEA and ammonia. *Energy Proc,.* **63:** 6590–6601.

Techno-economic evaluation of an ammonia-based post-combustion process integrated with a state-of-the-art coal-fired power plant. *Int. J. Greenh. Gas Control.* **31:** 87–95.

Jin, H., Wood, A. and Seiler, A. (2015). *Roadmap for carbon capture and storage demonstration and deployment in the People's Republic of China.* Asian Development Bank.

Kim, C. (2021). A review of the deployment programs, impact, and barriers of renewable energy policies in Korea. *Renew. Sust. Energy Rev.* **144:** 110870.

Knoope, M.M.J., Ramírez, A. and Faaij, A.P.C. (2013). A state-of-the-art review of techno-economic models predicting the costs of CO_2 pipeline transport. *Int. J. Greenh. Gas Control.* **16:** 241–270.

Korsbakken, J.I., Peters, G.P. and Andrew, R.M. (2016). Uncertainties around reductions in China's coal use and CO_2 emissions. *Nat. Clim. Change.* **6(7):** 687–690.

Kuckshinrichs, W. and Hake, J.-F. (2016). *Carbon Capture, Storage and Use: Technical, Economic, Environmental and Societal Perspectives.* Springer.

Lassagne, O., Iliuta, M.C., Gosselin, L. and Désilets, M. (2016). Techno-economic assessment of CO_2 capture from aluminum smelter emissions using PZ activated AMP solutions. *Can. J, Chem. Eng.* **94(4):** 761–770.

Li, M., Zhang, Q., Kurokawa, J.I., Woo, J.H., He, K.B., Lu, Z., Ohara, T., Song, Y., Streets, D.G., Carmichael, G.R. and Cheng, Y.F. (2015). MIX: A mosaic Asian anthropogenic emission inventory for the MICS-Asia and the HTAP projects. *Atmos. Chem. Phys. Discuss,* **15(23):** 34813–34869.

Li, L., Lin, J., Wu, N., Xie, S., Meng, C., Zheng, Y., Wang, X. and Zhao, Y. (2020). Review and outlook on the international renewable energy development. *Energy Built Environ.* **3:** 139–157

Getting ready for carbon capture and storage through a 'CCS (Carbon Capture and Storage) Ready Hub': A case study of Shenzhen city in Guangdong province, China. Energy. **36(10):** 5916–5924.

Systematic study of aqueous monoethanolamine (MEA)-based CO_2 capture process: Techno-economic assessment of the MEA process and its improvements. *Appl. Energy*. **165:** 648–659.

Technoeconomic assessment of an advanced aqueous ammonia-based postcombustion capture process integrated with a 650-MW coal-fired power station. *Environ. Sci. Technol.* **50(19):** 10746–10755.

Assessing the value of retrofitting cement plants for carbon capture: A case study of a cement plant in Guangdong, China. *Energy Convers. Manag.* **64:** 454–465.

Metz, B., Davidson, O., De Coninck, H.C., Loos, M. and Meyer, L. (2005). *IPCC special report on carbon dioxide capture and storage.* Cambridge: Cambridge University Press.

Mutezo, G. and Mulopo, J. (2021) A review of Africa's transition from fossil fuels to renewable energy using circular economy principles. *Renew. Sust. Energy Rev.* **137:** 110609.

NETL. (2021). *Carbon Storage Faqs.* netl.doe.gov. https://netl.doe.gov/coal/carbon-storage/faqs/carbon-storage-faqs (accessed November 30, 2021).

NREL. *Techno-economic analysis.* NREL.gov. https://www.nrel.gov/analysis/techno-economic.html (accessed November 29, 2021).

Nwaoha, C. and Tontiwachwuthikul, P. (2019). Carbon dioxide capture from pulp mill using 2-amino-2-methyl-1-propanol and monoethanolamine blend: Techno-economic assessment of advanced process configuration. *Appl. Energy*. **250:** 1202–1216.

Nwaoha, C., Beaulieu, M., Tontiwachwuthikul, P. and Gibson, M.D. (2018). Techno-economic analysis of CO_2 capture from a 1.2 million MTPA cement plant using AMP-PZ-MEA blend. *Int. J. Greenh. Gas Control*. **78:** 400–412.

Osagie, E., Biliyok, C., Di Lorenzo, G., Hanak, D.P. and Manovic, V. (2018). Techno-economic evaluation of the 2-amino-2-methyl-1-propanol (AMP) process for CO_2 capture from natural gas combined cycle power plant. *Int. J. Greenh, Gas Control*. **70:** 45–56.

Rao, A.B. and Rubin, E.S. (2002). A technical, economic, and environmental assessment of amine-based CO_2 capture technology for power plant greenhouse gas control. *Environ. Sci, Technol.* **36(20):** 4467–4475.

Riahi, K., Rubin, E.S., Taylor, M.R., Schrattenholzer, L. and Hounshell, D. (2004). Technological learning for carbon capture and sequestration technologies. *Energy Econ.* **26(4):** 539–564.

Roussanaly, S. and Grimstad, A.A. (2014). The economic value of CO_2 for EOR applications. *Energy Proc.* **63:** 7836–7843.

Roussanaly, S., Bureau-Cauchois, G. and Husebye, J. (2013). Costs benchmark of CO_2 transport technologies for a group of various size industries. *Int. J. Greenh. Gas Control*. **12:** 341–350.

Rubin, E.S. (2012). Understanding the pitfalls of CCS cost estimates. *Int. J. Greenh. Gas Control.* **10:** 181–190.

Rubin, E.S. (2019). Improving cost estimates for advanced low-carbon power plants. *Int. J. Greenh. Gas Control*. **88:** 1–9.

Rubin, E.S. and Zhai, H. (2012). The cost of carbon capture and storage for natural gas combined cycle power plants. *Environ. Sci. Technol.* **46(6):** 3076–3084.

Rubin, E.S., Mantripragada, H., Marks, A., Versteeg, P. and Kitchin, J. (2012). The outlook for improved carbon capture technology. *Prog. Energy Combust. Sci.* **38(5):** 630–671.

Van der Spek, M., Fout, T., Garcia, M., Kuncheekanna, V.N., Matuszewski, M., McCoy, S., Morgan, J., Nazir, S.M., Ramirez, A., Roussanaly, S. and Rubin, E.S. (2020). Uncertainty analysis in the techno-economic assessment of CO_2 capture and storage technologies. critical review and guidelines for use. *Int. J. Greenh. Gas Control*. **100:** 103113.

Stromberg, L. (2008). Pilot project update-focus a European projects. *Carbon Capture J.* **2:** 14.

Suman, A. (2021). Role of renewable energy technologies in climate change adaptation and mitigation: A brief review from Nepal. *Renew. Sust. Energy Rev.* **151:** 111524.

Tsupari, E., Kärki, J., Arasto, A. and Pisilä, E. (2013). Post-combustion capture of CO_2 at an integrated steel mill–part II: Economic feasibility. *Int. J. Greenh. Gas Control*. **16:** 278–286.

Van der Spek, M. and Ramirez, A. (2014). A structured approach for selecting carbon capture process models. a case study on monoethanolamine. *Energy Proc.* **63:** 1287–1295.

Van Vuuren, D.P., Hoogwijk, M., Barker, T., Riahi, K., Boeters, S., Chateau, J., Scrieciu, S., van Vliet, J., Masui, T., Blok, K. and Blomen, E. (2009). Comparison of top-down and bottom-up estimates of sectoral and regional greenhouse gas emission reduction potentials. *Energy Policy*. **37(12):** 5125–5139.

VITO. *Techno-Economic Assessment*. https://vito.be/en/product/techno-economic-assessment (accessed November 29, 2021).

Wara, M. (2007). Is the global carbon market working? *Nature.* **445(7128):** 595–596.

Wene, C.O. (2008). Energy technology learning through deployment in competitive markets. *Eng. Econ.* **53(4):** 340–364.

Wikipedia. *Techno-Economic Assessment* (Electronically published November 3, 2021). https://en.wikipedia.org/wiki/Techno-economic_assessment (accessed November 29, 2021).

Wu, N., Parsons, J.E. and Polenske, K.R. (2013). The impact of future carbon prices on CCS investment for power generation in China. *Energy Policy*. **54:** 160–172.

Yang, X., Pang, J., Teng, F., Gong, R. and Springer, C. (2021). The environmental co-benefit and economic impact of China's low-carbon pathways: Evidence from linking bottom-up and top-down models. *Renew. Sust. Energy Rev.* **136:** 110438.

Zhang, Z. (2015). Techno-economic assessment of carbon capture and storage facilities coupled to coal-fired power plants. *Energy Environ.* **26(6–7):** 1069–1080.

Zhang, D., Xu, M. and Li, A. (2012). Techno-economic analysis of a 300 MWe coal-fired power plant retrofitted with post-combustion CO_2 capture. *Asia-Pac J. Chem. Eng.* **7:** S201–S208.

The Current Methods and Practices of CO_2 Capture

Saber Kh. Elmabrouk and Mian Umer Shafiq

1 Introduction

According to the Global Carbon Project (2021), CO_2 emissions from fuel combustion, which had been stable for three years, began to rise again in 2017, reaching 32.8 billion tons. Data indicates faster growth in 2018. The report states that global fossil CO_2 emissions rose by 0.9% annually from 1990 to 1999 and by 3.0% annually from 2000 to 2009. Following the 2010 global financial crisis, emissions dropped by 1.2%, then resumed a 0.9% annual increase from 2010 to 2019, before plummeting by 5.4% in 2020 due to the COVID-19 pandemic, marking the largest decline ever, nearly five times the decrease seen in 2009 after the financial crisis. This is illustrated in Fig. 1. According to the IEA's 2022 analysis, emissions surged by 6% to 36.3 billion tons in 2021, the highest level recorded, as the global economy rebounded strongly from the pandemic and heavily relied on coal for energy production.

New coal records were reached in 2021 across Asia as energy demand increased, according to a 2022 report by the Global Energy Review (IEA 2022). These countries included Mongolia (+13%), India (+11%), China (+9%), Pakistan (+8%), the Philippines (+8%), and Kazakhstan (+6%). Coal power saw a significant increase in the US, the EU, and Japan in 2021 compared to 2020, although it was still below 2019 levels. China accounted for 54% of the world's coal power generation in 2021, up from 50% in 2019. Moreover, Fig. 2 displays the global power generation share by source during 1973 and 2019. This demonstrates that from 12.2% in 1973 to 23.6% in 2019, the use of natural gas (NG) as a source of power generation (reduced emissions) has increased. However, from 38.3% in 1973 to 36.7% in 2019, the usage of coal as the source of electricity generation (highest emission) dropped. Conversely, there has been a noticeable rise in the usage of RE (zero emissions) sources.

The COVID-19 epidemic caused a drop in CO_2 emissions, but the jump in worldwide CO_2 emissions of almost two billion tons was the highest absolute increase in history, according to an IEA study (IEA 2022). Unfavourable weather and energy market conditions, particularly a spike in natural gas prices, contributed to the 2021 jump in energy demand and increased coal use despite a record gain in RE generation. Continual efforts strive to reduce CO_2 emissions by employing various tactics to tackle climate change worldwide. This chapter describes a variety of techniques and procedures for CO_2 capture and reduction in the fossil fuel (coal, oil, and natural gas) electricity supply chain.

2 Electricity Supply Chain Measures

The electricity-generating plant, the transmission and distribution (T&D) system, and the final customer make up the power supply chain (Fig. 3). The power plant that generates energy using fossil fuels is the primary source of CO_2 emissions. Higher energy production within the generating plant would result from excessive end-user demand and losses in the T&D system, which would subsequently raise CO_2 emissions.

However, fossil fuels are the primary source of energy in traditional power plants. They are burned using three main technologies: (1) steam turbine (subcritical boiler), where heat produces high-pressure steam for electricity generation, as shown in Fig. 4; (2) gas turbine (combustion), where hot

S. Kh. Elmabrouk (✉)
Chemical and Petroleum Engineering, School of Applied Sciences and Engineering, Libyan Academy for Postgraduate Studies, Janzur, Libya
e-mail: saberelmabrouk@gmail.com

Department of Engineering Management, Faculty of Engineering, University of Tripoli, Tripoli, Libya

M. U. Shafiq
Chemical and Petroleum Engineering, Faculty of Engineering Technology and Built Environment, UCSI University Kuala Lumpur, Kuala Lumpur, Malaysia

J. Marrugo-Negrete et al. (eds.), *Remediation and Carbon Capture in Mining Sector*, https://doi.org/10.1007/978-3-031-87883-1_9

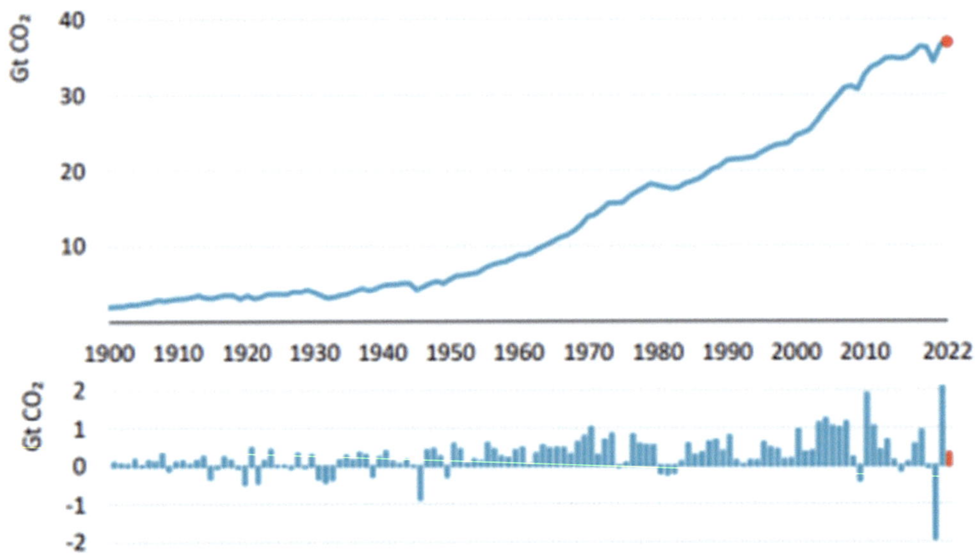

Fig. 1 Global fossil CO_2 emissions, 1900–2021

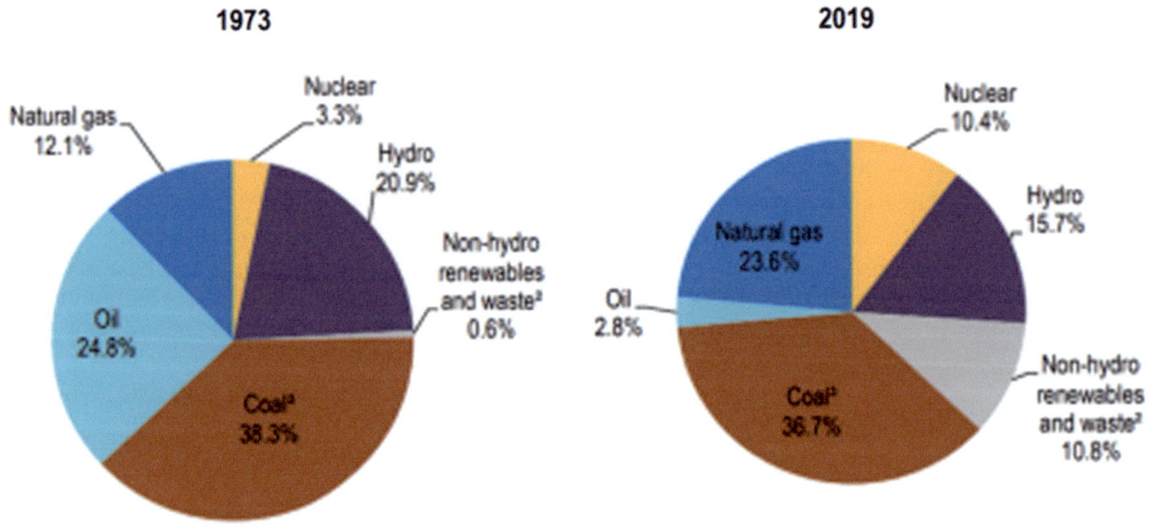

Fig. 2 Share of world electricity generation by source, 1973 and 2019 (IEA, September 2021)

Fig. 3 Electricity supply chain (Elmabrouke et al. 2017)

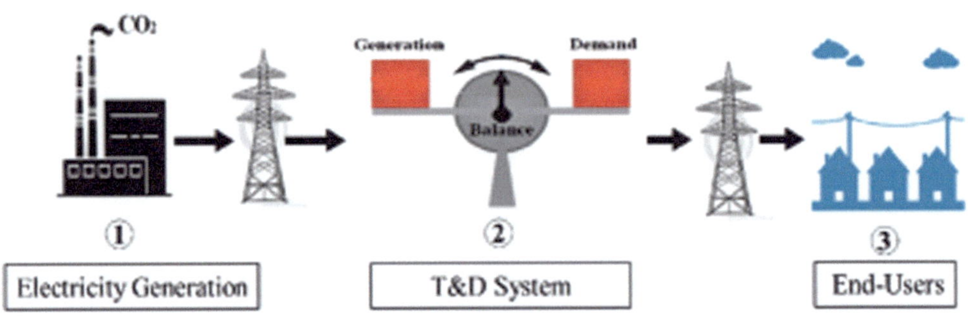

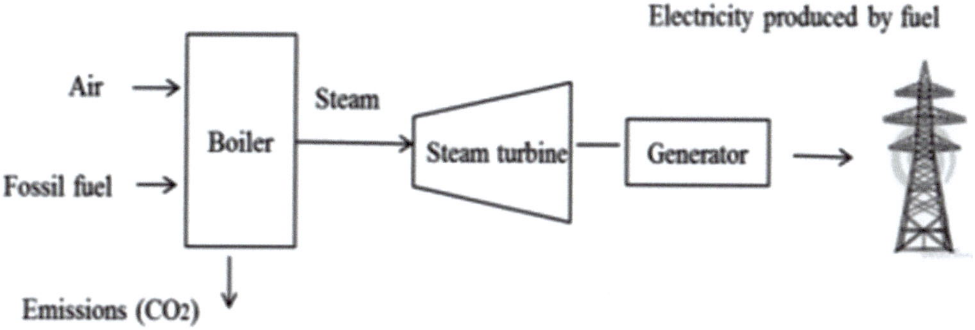

Fig. 4 Steam Turbine (Elmabrouke et al. 2017)

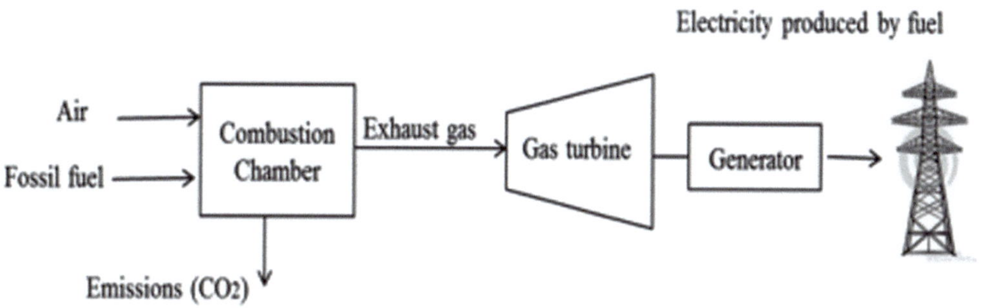

Fig. 5 Gas turbine, combustion (Elmabrouke et al. 2017)

exhaust gases pass through a turbine for electricity production, as shown in Fig. 5; and (3) combined cycle, which first burns fuel in a combustion turbine, utilises heated exhaust gases to generate electricity, and then uses the recovered gases to produce steam for additional electricity generation without extra fuel. This process, known as the Gas Turbine Combined Cycle (GTCC), is illustrated in Fig. 6. Additionally, Fig. 7 depicts the integrated coal gasification combined cycle (IGCC) technology, which incorporates the GTCC concept along with a coal gasifier.

On the other hand, Fig. 8 shows the three major elements of the electrical supply chain (the electricity-generating plant, the transmission and distribution (T&D) system, and the final customer) as well as potential CO_2 reduction strategies. There are three strategies to lower CO_2 emissions in the electricity-generating plant: (a) using non-fossil fuels, converting to low-carbon fuels, (b) improving power plant efficiency, and (c) turning on carbon capture power plants. In relation to the T&D system, the following actions can be taken: (a) EDM (energy data management); (b) improving system efficiency; and (c)

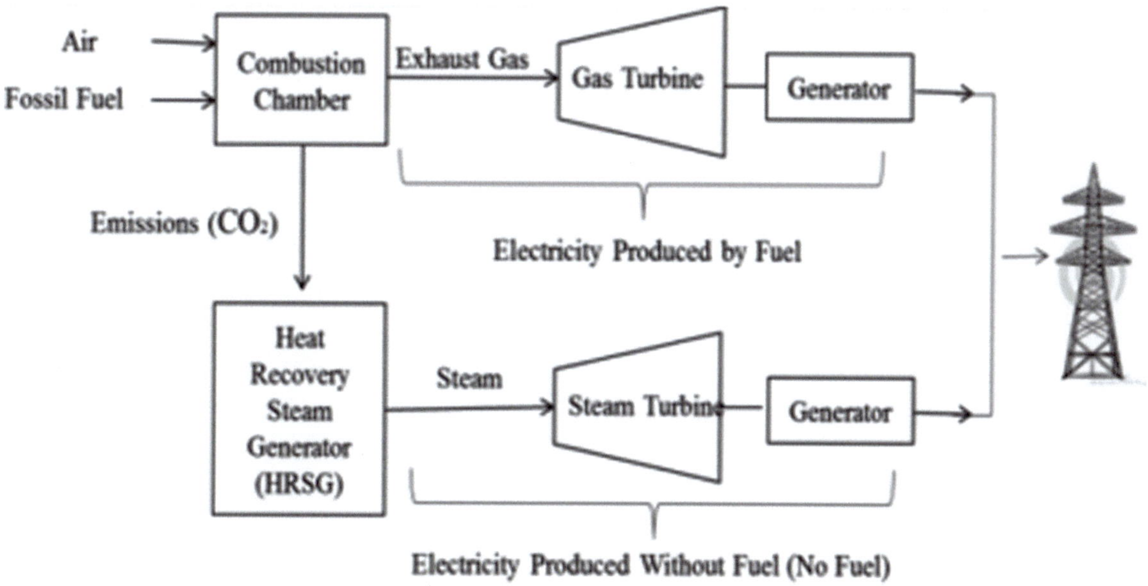

Fig. 6 Gas turbine combine cycle (Elmabrouke et al. 2017)

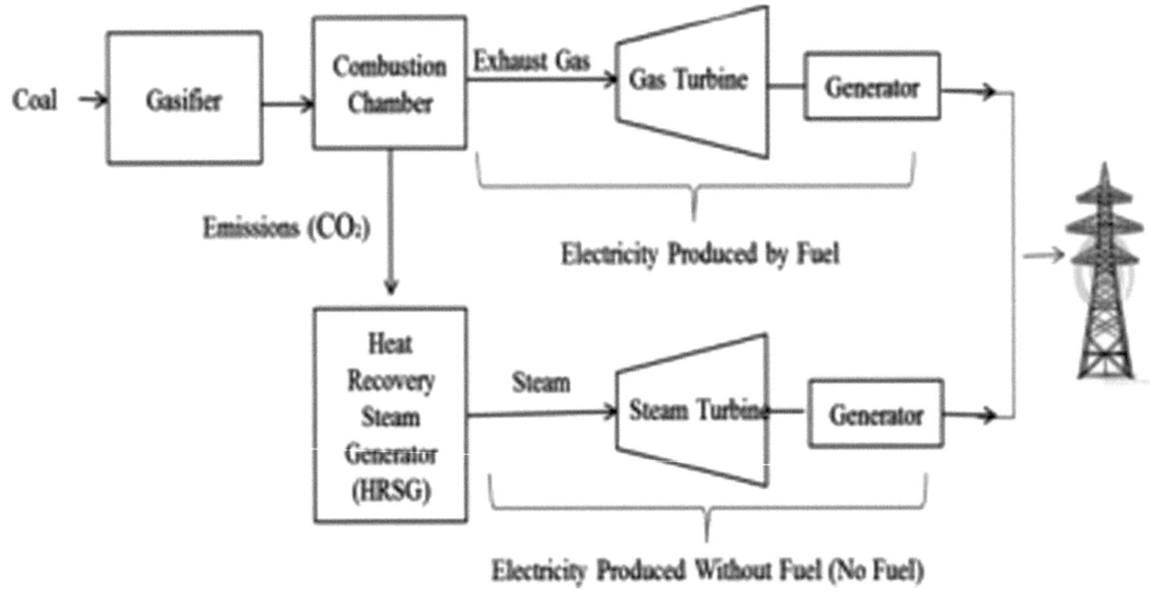

Fig. 7 Integrated coal gasification combine cycle (Elmabrouke et al. 2017)

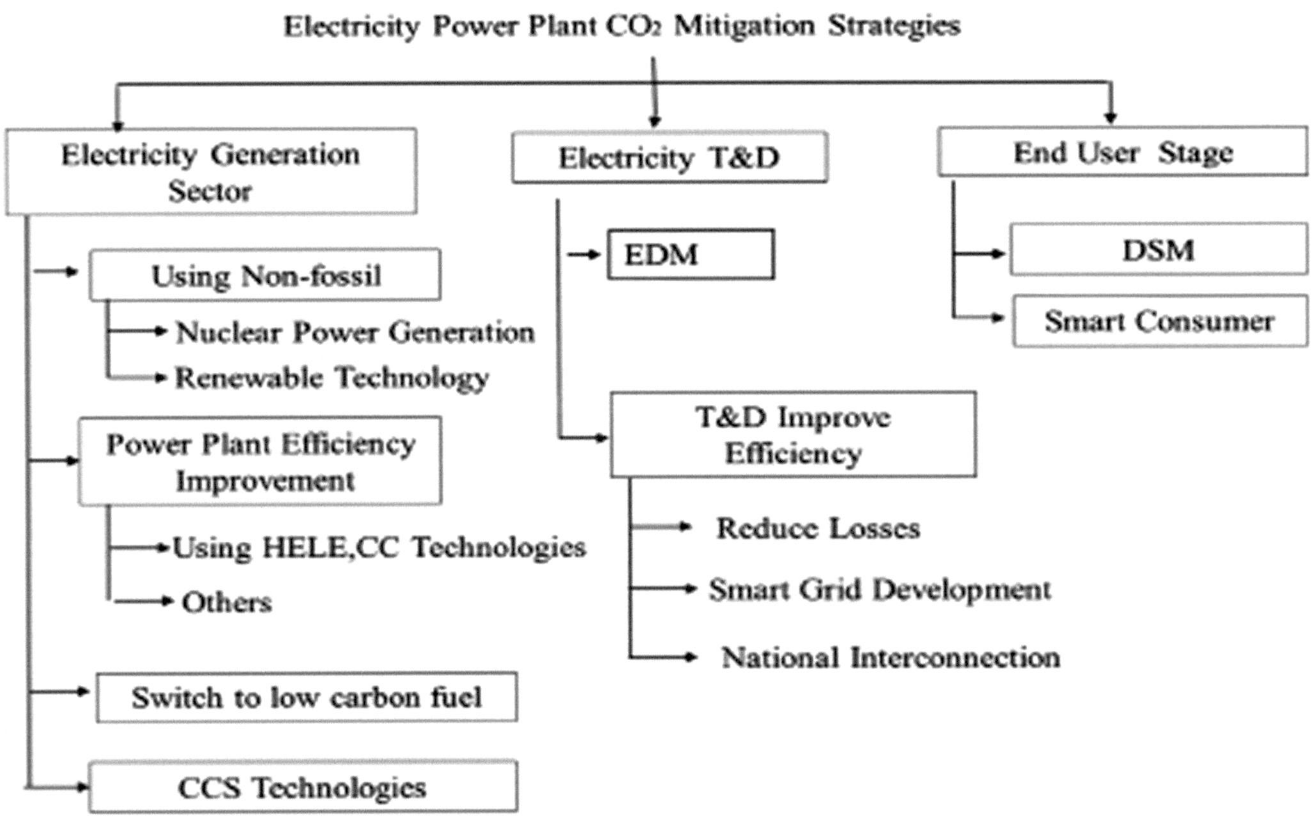

Fig. 8 CO_2 emissions mitigation strategies in the power generation supply chain

creating a smart grid. End-user activities that could be taken to cut CO_2 emissions can be summarised by demand-side management (DSM) and intelligent consumer behaviour.

2.1 Measures at the Electrical Generation Plant

The electrical generation plant is the stage where the electricity is generated. Four mitigation strategies can be applied; (1) Using Non-Fossil Fuel, (2) Power Plant Efficiency Improvement, (3) Switch to Low-Carbon Fuel, and (4) Carbon Capture and Storage (CCS) Strategy.

(1) Utilization of Non-fossil Fuels

Fossil fuels include coal, oil, and natural gas, while solar, wind, and hydropower are examples of non-fossil fuels, as well as nuclear power generation. However, as seen in Fig. 2, non-fossil fuel power production technologies accounted for 24.8% of total power generation in 1973 and 36.9% in 2019.

(a) Nuclear Power Generation

Nuclear reactors generate thermal energy by splitting uranium atoms, or the nuclear fission of an atom's nucleus, as the fundamental working principle of a nuclear power plant. With the help of this heat, the reactor turns water into steam at a high pressure and temperature, which powers a turbine generator to produce energy. During operation, nuclear power plants emit almost no greenhouse gases or air pollutants, and during their whole life cycle, they only emit very low amounts of emissions. According to Nuclear Power and Climate Change, using nuclear power prevents the annual release of more than 2 billion tons of CO_2. In 2018, 12 nations derived at least 25% of their electricity from nuclear power. Around 75% of France's electricity comes from nuclear power; more than 50% comes from Hungary, Slovakia, and Ukraine; and at least 30% comes from Belgium, Sweden, Slovenia, Bulgaria, Switzerland, Finland, and the Czech Republic. Currently, over 25% of Japan's electricity comes from nuclear power, and that percentage is predicted to climb (IAEA 2016).

Nuclear power is a very advanced technology utilised to create electricity due to its wonderful benefits, which include nearly zero CO_2 emissions. However, it has also encountered challenges, chiefly related to matters of public safety and notable hazards. (Daiet al. 2012). Regrettably, there are hazards and challenges linked to the expanding use of nuclear energy. These include negative public perception, operational risks and attendant safety concerns, uranium mining risks, financial and regulatory risks, waste management issues, and concerns about nuclear proliferation. Table 1 shows some examples of the world's nuclear power plant accidents according to the scale of international nuclear and rational events (The Rogers 2011).

(b) Renewable Power Generation

In 2018, IEA show that the first source of renewable electricity in the world is hydropower producing 13.3% and the second electricity sources is wind presents 8.9 % then comes bio-power and solar in the third place about 2%. Hydropower is expected to remain the world's primary source of renewable electricity generation and will be key to reducing carbon emissions from the power system while increasing system resilience (IEA 2016). As shown in Table 2, solar energy can be used for both solar heating and solar electricity.

Solar heat is one application for solar energy. Almost 398 GW of cumulative solar PV capacity and over 460 TWh of

Table 1 Nuclear power plant accidents

Country	Year	INES level*	IAEA description
Japan	2011	5	The reactor shutdown following the 2011 Sendai earthquake and tsunami resulted in an explosion due to an emergency cooling system malfunction
Ukraine	1986	7	Extensive consequences for the environment and health. A large portion of the reactor core inventory is being released externally
US	1979	5	Severe damage to the reactor core
Russia	1957	6	Due to a high-activity waste tank blowing up, a significant amount of radioactive material leaked into the environment
UK	1957	5	As a result of a reactor core fire, radioactive substances were released into the environment
Canada	1952	5	At AECL's NRX reactor, a reactor shutdown rod failure and multiple operator mistakes resulted in a significant power excursion of more than double the reactor's rated output

* 5: Accident with Wider Consequences, 6: Serious Accident, and 7: Major Accident

Table 2 Classification of applied solar energy

Solar/Sun energy application					
Solar electricity power			Solar heating		
PV electricity		Concentrated solar	Solar space heating	Solar water heating	
On-Grid	Off-Grid				

energy were produced in 2017, accounting for around 2% of global power output. Utility-scale projects account for little more than 60% of total PV installed capacity; distributed applications (off-grid, residential, and commercial) use the remaining portion. With a projected expansion of over 580 GW, solar PV is expected to lead the increase in renewable electricity capacity over the next 5 years, according to Ologunfemi et al. (2022). Regretfully, as Table 2 outlines, there are a few challenges and barriers when utilising RE.g or solar electricity. Table 3 outlines the barriers and challenges identified by Moorthy et al. (2019) in utilising RE.

(2) Improve Efficiency of the Electrical Power Plant
Operating at a lower efficiency means that a relatively large amount of fossil fuel is burned. As fossil fuel consumption increases, emissions also increase. So, improving the efficiency of power plants is one strategy to reduce the use of fossil fuels and subsequently to reduce CO_2 emissions. The efficiency of energy production can be improved with more efficient technology, the high-efficiency-low-emissions (HELE) technology.

To turn water into steam more quickly, HELE power plants run at higher air pressure and temperatures. Boilers and turbines become much more efficient as a result, saving fuel and cutting CO_2 emissions by up to 50%. There are several HELE technologies, including integrated gasification combined cycle, advanced ultra-supercritical, supercritical, and ultra-supercritical, that can be used to generate power with different levels of emissions and conversion efficiency. However, using HELE technology to increase the global average efficacy rate from 33 to 40% would cut CO_2 emissions by 2 billion tonnes annually, or the equivalent of India's annual emissions. (Wiatros-Motyka 2016).

(3) Switching to Low-Carbon Fuels (Natural Gas)
One of the main sources of energy in the world is natural gas, and its use is expanding quickly. In 2018, the rise in the world's overall energy demand was nearly entirely due to natural gas. In the energy industry, natural gas is used in many ways. It minimises air pollution and limits CO_2 emissions, and it can replace more polluting fuels. The World Energy Outlook team report from the IEA in 2019 offers supporting data for this crucial discussion. It mentions that switching to natural gas has already assisted in controlling the rise in worldwide emissions since 2010.

(4) Carbon Capture and Storage (CCS) Strategy
According to the Carbon Capture and Storage Association (2011), up to 90% of the CO_2 emissions caused by burning fossil fuels for industrial processes and energy production can be captured by using this CCS technique. Elmabrouke et al. (2017) presented that the CCS strategy consists of three stages; (1) *Capture/Separation Stage*, (2) *Transport Stage*, and (3) *Utilization/elimination Stage*. In the capture/separation stage, they listed three main types of capture system utilised in fossil fuel power system; *post-combustion, pre-combustion,* and *oxy-combustion.* According to Kanniche et al., there are three methods of capture: (1) pre-combustion capture, which involves capturing CO_2 in a synthesis gas after CO is converted to CO_2; (2) post-combustion capture, which involves collecting CO_2 from exhaust gases after fuel has burned completely in the presence of air; and (3) capture in oxy-combustion, which entails burning fuel in oxygen while recycling exhaust gases (which are primarily made up of CO_2 and water) and purifying the CO_2 flow to remove incondensable gases.

According to Natural Energy Technology Laboratory, (n.d.), Table 4 lists some CCS power plants operation projects. The table shows that the majority of CCS projects use the capture system in post-combustion, where CO_2 is captured in the exhaust after the fuel has been completely burned with air.

2.2 Measures Taken in T&D System

The power grid transfers electricity from the electric power plant to the final consumers once it has been generated. The overhead wires and ground cables that make up the so-called T&D system are this power grid. Consequently, the T&D system plays a crucial role in achieving the benefits of minimising the effects of the two electrical supply chain components, acting as more than just a physical bridge connecting the generation sector to the end-user phase. Due to equipment resistance and line loss in the cables, a portion of the generator's energy is lost during the transmission process. Heat is produced from this energy mostly. The T&D system's

Table 3 Barrier and challenges in the using of RE

Barriers		Description	References
Social Barriers	Lack of awareness of the benefits	Inadequate awareness of RE technology, insufficient data on the financial and ecological benefits, and doubts over the sustainability of RE projects financially	Nasirov et al. (2015)
	Disruption of seascape	The RE projects are viewed by most as environmentally degrading, and the lack of community consultation results in public opposition, political leaders, community organizations, and, in some cases, environmental organizations	Nasirov et al. (2015)
	Loss of alternative income	One issue with wind and solar farms is the large land requirement compared to coal plants. Developing RE plants on a large scale involves converting agricultural or marine areas, leading to conflicts, environmental worries, and ecosystem disruptions. Balancing RE needs with land and wildlife preservation is a complex challenge	Chauhan and Saini (2015), Nesamalar et al. (2017)
	Lack of professionals	A strong base of workers and experts with knowledge is needed to plan, build, operate, and maintain RE applications. Lack of trained technical professionals and lack of training institutions hinder RE technologies from expanding	Ansari et al. (2016)
Economic Barriers	Fierce competition with fossil fuels	Fossil fuels—especially coal—will remain essential to the production of energy. In most countries, coal remains the most popular fuel due to its accessibility, affordability, and abundance. Fossil fuels are expected to provide 78% of the world's energy by 2040	Dulal et al. (2013)
	Government grants and subsidies	The majority of nations provide significantly more public subsidies for conventional energy than they do for RE. For instance, government funding for mining and exploration initiatives is still provided to coal businesses in Australia and Indonesia	Dulal et al. (2013)
	Fewer financing nstitutions	Initiatives for financing RE are not well funded or supported by a large number of groups. For this reason, it is very challenging for RE developers and producers to secure funding for projects that have interest rates that are similar to those of fossil fuel facilities	Ansari et al. (2016), Ohunakin et al. (2014)
	High initial capital cost	Investors are reluctant to enter the RE projects, especially solar energy, due to the high initial capital cost and long payback period	Ansari et al. (2016), Suzuki (2013), Painuly (2001)
	Intangible costs	The expenses of exploration, production, distribution, and usage are all included in the overall cost of fuel in practically all nations, but they do not include the costs of damage to the environment and effects on health and the atmosphere	Arnold (2015)
Technological Barriers	Limited availability of infrastructure and facilities	It is not common to find advanced RE technologies, especially in developing countries. Even if this technology were widely available, it would be quite costly to purchase, considering that the majority of RE installations are located in remote locations. Since most existing networks were not designed to be connected with the RE, they will most likely need to be updated or changed	Dulal et al. (2013), Izadbakhsh et al. (2015)
	Absence of an operation and maintenance culture	Because RE technology is still in its infancy, there is a dearth of experience with operations and maintenance. The absence of equipment and spare parts in the area also raises production expenses dramatically, which raises the overall cost	Sen and Bhattacharyya (2014), Bhandari et al. (2015)
	Lack of R&D capabilities	R&D spending is insufficient for RE to be economically competitive with fossil fuels. Because RE is still in its early stages of development and has a significant risk, both governments and energy businesses are reluctant to spend on R&D	Cho et al. (2013)
	Technology complexities	The lack of adequate protocols, rules and standards for robustness, reliability, performance, etc. in RE technologies prevents the widespread commercialization of RE. Energy storage is also one of the major technological barriers today	Nasirov et al. (2015)

(continued)

Table 3 (continued)

Barriers		Description	References
Regulatory Barriers	Ineffective policies by government	Strict regulation of the energy sector is essential both for sustainable development and for resolving the conflict between renewable and non-renewable resources. When it comes to grant implementation, various agencies are confused due to the lack of effective policies and policies to integrate RE technology with the global market, which hampers RE projects	Zhang et al. (2014)
	Inadequate fiscal incentives	Governments subsidize RE sources through tariffs in an attempt to make them more affordable than technologies reliant on fossil fuels. However, government actions to eliminate tariffs on imports of the machinery and parts needed for RE facilities have not been sufficient. In the event that these financial incentives are not sufficient, high costs arise that impede the growth, upkeep, and operation of the industry and set it back for future development	Sun and Nie (2015)
	Administrative and bureaucratic complexities	The implementation of RE projects is hampered by a variety of factors, including but not limited to administrative ones such as restrictions and delays in planning. The lengthy lead time for acquiring licenses and the lack of coordination between several authoriti es needlessly lengthen the project's development period	Ahlborg and Hammar (2014)
	Unfeasible government promises	There is a gap between the actual outcomes of RE project implementation and the policy objectives declared by some governments. Therefore, policies need to be formulated that offer clear guidance on key legislative and regulatory matters to foster a stable and growth-oriented industry	
	Lack of standards and certification	To ensure that machinery and parts manufactured or procured abroad meet the needs of the importing company, standards and certifications are essential. These certifications ensure that companies are utilizing RE plants in accordance with local laws. For energy producers, the absence of such standards leads to uncertainty and unnecessary challenges	Emodi et al. (2014)

Table 4 CCS power plants operating projects

Country	Project name	Project start date	Post-	Pre-	Oxy-	CO_2 captured T/d	CO_2 capture technology	References
Australia	H3 Capture Project—CO_2 CRC	2009	✓			25	Solvent	Qader et al. (2011)
	Mulgrave Capture Project—CO2 CRC	2009		✓		741	Amine Ammonia Membrane	Anderson et al. (2011)
	E.ON Coal Plant in Wilhelmshaven	2010	✓			70	Fluor Econamine	Reddy et al. (2013)
	E.ON Pilot Plant in Heyden	2010	✓			90	Amines	
Germany	RWE Power AG Lignite Plant	2009	✓			2,000	Amines	Renzenbrink et al. (2009)
Japan	Matsushima Coal Plant Capture Project	2006	V			10	Amines	Kishimoto et al. (2009)
Spain	Puertollano Plant	2010		✓		100	Pressurized Entrained Flow	Casero et al. (2013)
Sweden	Sargas Vartan	2007	✓				Solvent	Bryngelsson et al. (2009)
USA	Mitchell Energy Bridgeport Plan	1991	✓			500	Amino Acid Salt-Based Solvent	
	Pleasant Prairie Pilot Plant	2008	✓			81.1	Chilled Ammonia	Kozak et al. (2009)

physical properties have a significant impact on how much energy is lost in the system (Gadeppanavar et al. 2013).

CO_2 reduction measures in the electrical T&D system can be achieved by:

(1) *Energy Data Management (EDM) Techniques.* EDM provides collected data to balance between power generation and end-users.
(2) *Smart Grid Technology.* In addition to serving as a physical link between the customer (end-user) and power producing systems, smart grid technology is crucial for delivering the benefits of system mitigation. The smart grid is thought to be a means of lowering energy consumption and increasing the effectiveness of the electrical network by connecting power produced by fossil fuels and non-fossil fuels to the grid.
(3) *Improving Grid Efficiency* by reducing losses. The reduction of distribution losses represents a specific issue for each Distribution System Operator, not only due to heterogeneous levels of both technical losses and non-technical losses but also due to differences on definition, measurement or regulation of losses.
(4) *National Interconnection.* The main advantage of national interconnection is to postpone or avoid the construction of new power plants (reducing CO_2 emissions). Different countries share electricity by interconnected networks without affecting their safety and reliability. Additionally, interconnection decreases the requirement for backup capacity to handle variations in demand, which lowers operating expenses and emissions.

2.3 Measures Taken in End-Users Stage

The best way to lower CO_2 emissions at the end-user level is to save energy. Residential, commercial, and industrial consumers are the end-users of electricity. Reduced electricity use can lower CO_2 emissions for all kinds of end-users. This can be accomplished by reducing the amount of electricity used, and all end-users can become wise consumers.

2.4 CO_2 Separation Technologies

Different types of CO_2 separation technologies have been utilised to separate CO_2 from gas mixtures. Various technical approaches and materials for each technique are mentioned in Fig. 9, except for the cryogenic separation technique. Mostly, the focus of research related to carbon capture is based on mature technologies like chemical absorption which is done using amine-based solvents. But, the solvents used in large volumes require a significant amount of thermal energy for regeneration. Due to this disadvantage, research is shifted towards emerging techniques like calcium looping, membranes, algae-based capture, sorbents, and liquefaction (Song et al. 2019). However, the technology selected is based on the product requested and the conditions of the gas stream treated for example CO_2 pressure, CO_2 concentration, temperature, and the level and type of species (Audus 2000).

2.4.1 Absorption

This method of separation is divided into two categories: chemical absorption and physical absorption. The commercially most used technique is chemical absorption (Drage et al. 2007).

Physical Absorption

The solubility of CO_2 in the solvent controls the CO_2 removal through the physical absorption phenomenon, whereas CO_2 solubility is dependent on the partial pressure and feed gas temperature. Low temperature and higher partial pressure of CO_2 increase its solubility in the solvent which is being utilised as absorbents. Later, by increasing the temperature or by reducing the pressure, these solvents can be regenerated.

Physical absorption is thought to be an effective method for extracting CO_2 and H_2S (acid gas) from natural gas. CO_2 is also extracted from the synthesis gas, which is then utilised to create methanol, ammonia, and hydrogen. The solvents that have been explored the most are Selexol (polyethylene glycol dimethyl ether) (Johnson and Homme 1984), Rectisol process (Kovvali and Sarkar 2002) propylene carbonate (Flour process), and Purisol. The process is absorption is usually done when the partial pressure of CO_2 is high and the energy requirement is obtained from the pressurization of flue gases. Therefore, the process of physical absorption is not cost-effective for gas streams when partial pressure is less than 15 vol% (Chakravarti et al. 2001).

Chemical Absorption

In this technique, CO_2 can be captured even if the partial pressure of CO_2 is low. Therefore, this process is preferable when CO_2 needs to be captured during the post-combustion process in power plants. Because of the high selectivity of chemical absorption, pure CO_2 can be obtained. Thus, at industrial-scale CO_2 capture, chemical absorption is a suitable technique.

Solvents:

The ideal solvent must have high reactivity with the CO_2, less regeneration cost, high absorption capacity, high thermal stability, low impact on the environment, less solvent degradation, and low solvent cost (Davidson 2007). The chemical solvents are responsible for fast kinetics that capture from streams having a low partial pressure of CO_2.

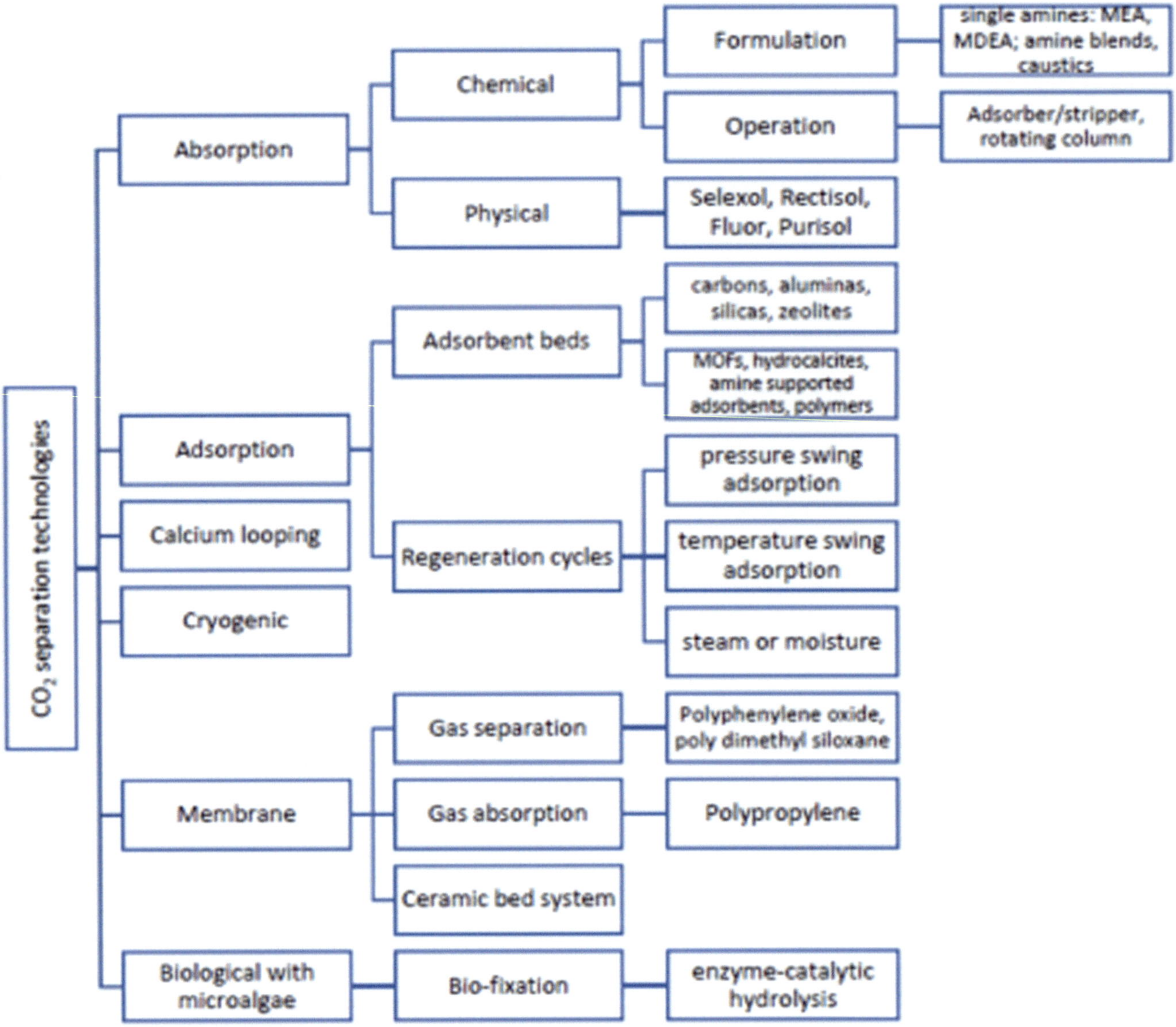

Fig. 9 CO_2 Separation technologies

Additionally, the process of wet scrubbing causes heat integration and ease of heat management which is very helpful in exothermic absorption reactions. Monoethanolamine (MEA) which is an amine-based scrubbing solvent (Bello and Idem 2005) and a solution of ammonia are two very important solvents (Idem et al. 2006). When MEA is used as a solvent, it can yield higher CO_2 capture capability and consume less energy.

Nevertheless, these solvents pose various types of thermodynamic, toxicity, and corrosivity weaknesses. Moreover, a large quantity of heat stream is required for the regeneration of solvents which de-rates power plants significantly. Additionally, a significant amount of heat energy is required to cool, heat, and pump a non-reactive carrier, which in most cases is water. The process of vacuum stripping can reduce the requirement of heat stream for regeneration of solvent but is quite expensive. When MEA is used in the operation of capture, the energy loss ranges from 15 to 37% of the plant's net power output (Herzog and Drake 1993). Other than raising the operational costs, this method produces a large amount of CO_2 which mitigates the benefit of sequestration.

To encounter these problems, new solvents have been studied by the researchers such as piperazine (new amine formulations), blends of amines, carbonates, as well as ionic liquids. They all have improved kinetics, smaller equipment size, and low regeneration energy requirements. Ionic liquids are the most recent solvents on which the MEA capture process has been utilised. Low volatility, high chemical and thermal stability, a broad range of liquid temperatures, and changeable physiochemical characteristics make it a unique solvent (Wappel et al. 2010). Therefore, to capture CO_2, the ionic liquid can be considered an alternative solvent.

Solid Sorbents:

Due to the disadvantages posed by liquid absorbents such as corrosion, solid absorbents have been introduced (Liu et al. 2006). Usually, sorbents like mesoporous materials that have a large surface area and pore volume are suitable. These sorbents redivided into two types, metal-based solvents (alkali) and amine-based sorbents like Na_2CO_3, CaO, and K_2CO_3 (Zhao et al. 2009). The amine-based solvents require less heat for regeneration compared to aqueous amines. However, the thermal stability and mechanical properties of solid-amine-based solvents are poor (Choi et el. 2009). Solid solvents have high temperatures compared to aqueous amines. Moreover, the handling of solid solvents is easier compared to aqueous amines. Different operational conditions for various types of solvents (Majchrowicz et al. 2009), the potential of sorbents to store CO_2 (Aronu et al. 2010), and CO_2 absorption kinetics in solid sorbents (Hoist et al. 2009) are some issues that still need to be discussed about solid sorbents.

2.4.2 Membrane-Based Separation

The contact between the liquid solvent and the gas stream is functioned by membranes. The membrane materials determine the cost and performance of these membrane-based technologies to capture CO_2 on a large scale. These membranes can be of two types: organic (polymeric) (Scholes 2009) and inorganic (zeolite, carbon, metallic, and ceramic) and they can be non-porous or porous (Hsieh 1990). Department of Energy (DOE) in the US fabricated a new polymer membrane material with significant chemical stability. Membranes are useful when the partial pressure of CO_2 is above 20 vol% which is considered as high concentration CO_2 applications, e.g. flue gas streams obtained from oxyfuel and processes of integrated gasification (Favre 2007). Permeability of membrane (flux through the membrane), stability of chemical and selectivity (allow the flow of one gas while blocking the other) as well as high temperature (>150 °C) mechanical stability is used to determine the performance of the membrane. Membranes can work at high temperatures and pressures.

The most important usage of membrane technology is the removal of CO_2 and purification of gas during hydrogen production, but its utilization for pre-combustion CO_2 capture is still new. The membrane technology is more attractive compared to absorption or adsorption-based capture of CO_2 because it does not cause absorption and adsorption losses. It is fit for large-scale CO_2 applications because this membrane-based technology is compact and simple.

2.4.3 Adsorption

It is a CO_2 capture, one of the most promising technologies and it can be applied at an industrial-scale post-combustion CO_2 process (Drage et al. 2009). Moreover, the energy requirement is low and capital development cost is low, which ultimately helps in the development of CO_2 capture and storage. In this process, a selective adsorbent is used and CO_2 is captured on the surface. To optimise the separation process, various adsorbents (Zhao et al. 2007) can be combined with various process options. The quality of adsorption depends on the adsorbed particles' properties like molecular size, polarity, molecular weight, and properties of absorbent surface also like polarity and pore size. There are two types of adsorption named vacuum swing adsorption (VSA) that work at ambient conditions of temperature and pressure. The other one is pressure swing adsorption (PSA) which works at a higher pressure than atmospheric pressure. Various issues are related to the adsorption process like the material of sorbent (Sjostrom and Krutka 2010), design and operating conditions, and conditions of feed gas like pressure, impurities, temperature, and concentration (Zhang et al. 2008). Electric Swing Adsorption (ESA) and Temperature Swing Adsorption (TSA) can be used to regenerate the adsorbent by changing the temperature as adsorption is an exothermic process. To reduce the energy consumption of capture plants, an investigation of operating conditions has been performed (Tlili et al. 2009). The adsorption process requires less energy as compared to the absorption process (Damen et al. 2006). When adsorbents are used to capture CO_2 in the pre-combustion process, it helped to increase the yield of H_2. Adsorption is a physical phenomenon, therefore adsorbents like hydrotalcite (materials containing MG/Al/Zr), nanotubes (Thiruvenkatachari et al. 2009) fibre-phenolic resins of activated carbon, carbon fibre monolithic adsorbent (An et al. 2009), red mud, and amine immobilise adsorbents (Chatti et al. 2009) have been investigated to improve the quality of CO_2 removal. The stability of carbon-based materials is excellent compared to other adsorbents.

2.4.4 CO_2 Capture and Fixation Through Algal Biomass Production

Due to the presence of NO_x, SO_x and a high concentration of CO_2 (more than 50%), the growth of strains of green algae agitated (Hanagata 1992). These three gases are the major contaminants due to the burning of fossil fuels (Brown 1996). High-value biomolecule production increases when exposed to high concentrations of CO_2. Solar energy helps green algae to produce more hydrogen through the process of biophotolysis (Ghirardi et al. 2000). The maximum energy and economical efficiency from the algae process can be achieved if the CO_2 fixing capacity and solar energy conversion efficiency are high. These species can produce H_2 by converting solar energy and can also produce a high amount of biomolecules.

In actual practice also to produce H_2, green algae are grown by solar energy and by photosynthesis. Later, the sulphur may be removed from the algae. The retention time of fixed CO_2 is 50–100 years if it is fixed in agriculture (Grimston et al. 2001). Around 26 kg/h of CO_2 can be captured in a $100m^3$ bioreactor and tested in the laboratory (Cheng et al. 2006).

2.4.5 Chemical Looping Combustion (CLC)

CLC is a novel method to capture CO_2 with less thermal energy requirements compared to other techniques. The main advantage of chemical looping in the combustion process is the in situ creation of oxygen, which eliminates the demand to separate oxygen from the air and reduces cost and energy demands. When fossil fuels are burnt in the presence of pure oxygen, it allows for capturing CO_2 in power plant applications. The oxygen is separated from the air in the oxy-combustion power generation that provides oxygen to the combustion process. While, oxygen is produced internally in the chemical looping system which reduces energy, operating, and capital costs.

In Chemical Looping Combustion (CLC), the system receives oxygen via reduction–oxidation cycling of the oxygen carrier, which is usually a metal-based solid material like the oxides of iron, nickel, or copper. It may also be a metal oxide of alumina or silica (a high surface area substrate) that is usually not a part of the reaction. The processes of oxidation and reduction are split into separate reactors during the typical CLC process. The oxygen for combustion is supplied by the metal oxide, while, in the fuel reactor, it is reduced which is operating at increased temperatures. Based on the fuel and metal oxide carrying oxygen, the reaction can be both endothermic and exothermic. The product obtained from the combustion in the fuel reactor is concentrated CO_2. The water stream obtained can be compressed, purified, and can be stored for other uses. Then, the oxygen carrier which was reduced due to the reaction was transferred to the air reactor (working at a high temperature) and its oxidised state is regenerated. A hot gas stream is produced in the air reactor, which is then used to produce steam that can drive a turbine to generate power. This whole process is repeated once the oxygen carrier is returned to the fuel reactor. The current research is focused on developing the carriers than can carry sufficient oxygen, durability, and production cost to reduce the operating cost, develop solids separation and circulation techniques, and improve the design of the reactor and the system optimisation and design.

2.4.6 Cryogenic Carbon Capture Technologies

They are usually low-temperature carbon capture technologies. It relies on the phase change, therefore, CO_2 can be separated in the form of solid or liquid from the gas (Berstad et al. 2013). The processes that occurred when the pressure is less than 120 K are known as 'Cryogenics', for example, nitrogen and oxygen condensation. The processes where the temperature is above 120 K are known as conventional refrigeration (Timmerhaus and Reed 2007). However, it is usually known as low-temperature separation. The cryogenic separation process is not too much discussed because it is expensive. High-purity product is the main advantage of this technique. Moreover, toxic chemicals were not required and this technique can be applied to a range of CO_2 concentrations.

3 Conclusion

In conclusion, CO_2 emissions from fuel combustion have experienced fluctuations over the past few decades, with a significant drop in 2020 due to the COVID-19 pandemic, followed by a surge to the highest level on record in 2021. Efforts to reduce CO_2 emissions include various strategies such as utilising non-fossil fuels, improving the efficiency of power plants, switching to low-carbon fuels, and implementing carbon capture and storage (CCS) strategies. The use of nuclear power and renewable energy sources, such as hydropower and solar energy, are important components of the non-fossil fuel approach. Improving the efficiency of power plants, transitioning to low-carbon fuels like natural gas, and implementing CCS strategies can also contribute to emissions reductions. Various methods of CO_2 separation technologies, including absorption, membrane-based separation, adsorption, algal biomass production, chemical looping combustion, and cryogenic carbon capture technologies, offer potential for capturing and reducing CO_2 emissions, though each method comes with its own set of challenges and considerations. Continued research and development in these areas will be crucial for achieving global emissions reduction targets and addressing climate change.

References

Ahlborg, H., Hammar, L. (2014) Drivers and barriers to rural electrification in Tanzania and Mozambique—Grid-extension, off-grid, and renewable energy technologies. *Renewable Energy*, **61:** 117–124. https://doi.org/10.1016/j.renene.2012.09.057.

An, H., Feng, B. and Su, S. (2009). CO_2 capture capacities of activated carbon fiber-phenolic resin composites, *Carbon*, **47:** 2396–2405.

Anderson, C., Scholes, C., Lee, A., Smith, K., Kentish, S., Stevens, G., Webley, P.A., Qader, A., Hooper, B. (2011). Novel pre-combustion capture technologies in action–Results of the CO2CRC/HRL Mulgrave capture project. *Energy Procedia*, **4:** 1192–1198. https://doi.org/10.1016/j.egypro.2011.01.173.

Ansari, S., Garud, R., Kumaraswamy, A. (2016) The disruptor's dilemma: TiVo and the U.S. television ecosystem. *Strategic Management Journal*, **37(9):** 1829–1853. https://doi.org/10.1002/smj.2442.

Arnold, G. (2015). Street-level policy entrepreneurship. *Public Management Review*, **17(2):** 307–327.

Aronu, U.E., Svendsen, H.F. and Hoff, K.A. (2010). Investigation of amine amino acid salts for carbon dioxide absorption. *International Journal of Greenhouse Gas Control*, **4(5):** 771–775.
Audus, H. (2000). Leading options for the capture of CO_2 at power stations. Presented at the 5th international conference on greenhouse gas control technologies, Australia.
Bello, A. and Idem, R.O. (2005). Pathways for the formation of products of the oxidative degradation of CO_2-loaded concentrated aqueous monoethanolamine solutions during CO_2 absorption from flue gases. *Industrial & Engineering Chemistry Research,* **44(4):** 945–969.
Berstad, D., Anantharaman, R. and Nekså, P. (2013). Low-temperature CO_2 capture technologies—Applications and potential. *International Journal of Refrigeration*, **36:** 1403–1416.
Bhandari, K.P., Collier, J.M., Ellingson, R.J., Apul, D.S. 2015. Energy payback time (EPBT) and energy return on energy invested (EROI) of solar photovoltaic systems: A systematic review and meta-analysis. *Renewable and Sustainable Energy Reviews*, **47:** 133–141. https://doi.org/10.1016/j.rser.2015.02.057.
Brown, L.M. (1996). Uptake of carbon dioxide from flue gas by microalgae. *Energy Conversion and Management*, **37 (6–8):** 1363–1368.
Bryngelsson, M., Lindfeldt, E.G., Heller, J. (2009) The modelling of a hybrid combined cycle with pressurised fluidised bed combustion and CO_2 capture. *International Journal of Greenhouse Gas Control*, **3(3):** 255–262. https://doi.org/10.1016/j.ijggc.2008.09.002.
Casero, P., García-Peña, F., Coca, P. (2013). Elcogas Pre-Combustion Carbon Capture Pilot. Real Experience of Commercial Technology. *Energy Procedia*, **37:** 6374–6382. https://doi.org/10.1016/j.egypro.2013.06.567.
Chakravarti, S., Gupta, A. and Hunek, B. (2001). Advanced technology for the capture of carbon dioxide from flue gases. In 1st National Conference on Carbon Sequestration, Washington DC.
Chatti, R., Bansiwal, A.K., Those, J.A., Kumar, V., Jadhav, P., Lokhande, S.K., Biniwale, B., Labhsetwar, N.K., Rayalu, S.S. (2009). Amine loaded zeolites for carbon dioxide capture: amine loading and adsorption studies. *Microporous and Mesoporous Materials*, **121:** 84–89.
Chauhan, A., Saini, R.P. (2015). Renewable energy based off-grid rural electrification in Uttarakhand state of India: Technology options, modelling method, barriers and recommendations. *Renewable and Sustainable Energy Reviews*, **51:** 662–681. https://doi.org/10.1016/j.rser.2015.06.043.
Cheng, L.H., Zhang, L., Chen, H.L. and Gao, C.J. (2006). Carbon dioxide removal from air by microalgae cultured in a membrane-photobioreactor. *Separation and Purification Technology*, **50(3):** 324–329.
Cho, S., Crenshaw, K.W., McCall, L. (2013). Toward a field of intersectionality studies: Theory, applications and praxis. *Signs*, **38(4)**. https://doi.org/10.1086/669608.
Choi, S., Drese, J.H. and Jones, C.W. (2009). Adsorbent materials for carbon dioxide capture from large anthropogenic point sources, *Chem. Sus. Chem*, **2(9):** 796–854.
Dai, P., Chen, G., Zhou, H., Su, M. and Bao, H. (2012). CO_2 mitigation measures of power sector and its integrated optimization in China. *The Scientific World Journal*, 907685.
Damen, K., Van Troost, M., Faaij, A. and Turkenburg, A. (2006). A comparison of electricity and hydrogen production systems with CO_2 capture and storage. Part A: Review and selection of promising conversion and capture technologies. *Progress in Energy and Combustion Science*, **32:** 215–246.
Davidson, R.M. (2007). Post-combustion Carbon Capture from Coal-Fired Plants – Solvent Scrubbing. *IEA Clean Coal Centre*, CCC/125.
Drage, T.C., Arenillas, A., Smith, K.M., Pevida, C., Piipo, S. and Snape, C.E. (2007). Preparation of carbon dioxide adsorbents from the chemical activation of urea-formaldehyde and melamine-formaldehyde resins. *Fuel*, **86:** 22–31.
Drage, T.C., Smith, K.M., Pevida, C., Arenillas, A. and Snape, C.E. (2009). Development of adsorbent technologies for post-combustion CO_2 capture. *Energy Procedia*, **1:** 881–884.
Dulal, H.B., Shah, K.U., Sapkota, C., Uma, G., Kandel, B.R. (2013). Renewable energy diffusion in Asia: Can it happen without government support?, *Energy Policy*, **59:** 301–311. https://doi.org/10.1016/j.enpol.2013.03.040.
Elmabrouk, S.K., Bader, H.E. and Mahmud, W.M. (2017). An overview of power plant CCS and CO_2-EOR projects. In: *International Conference on Industrial Engineering and Operations Management*.
Emodi, V.N., Yusuf, S.D., Boo, K.J. (2014). The necessity of the development of standards for renewable energy technologies in Nigeria. *Smart Grid and Renewable Energy*, **5(11)**: 259.
Favre, E. (2007), Carbon dioxide recovery from post-combustion processes: Can gas permeation membranes compete with absorption. *Journal of Membrane Science*, **294:** 50–59.
Gadeppanavar, M. and Pattanashetti, V. (2013). Improving Efficiency in Transmission and Distribution System. *International Journal of Emerging Technology and Advanced Engineering*, **3(3):** 157–160
Ghirardi, M.L., Zhang, J.P., Lee, J.W., Flynn, T., Seibert, M., Greenbaum, E. and Melis, A. (2000). Microalgae: A green source of renewable H_2. *Trends in Biotechnology*, **18(12):** 506–511.
Global Carbon Project (2021). Supplemental data of Global Carbon Budget 2021 (Version 1.0) Global Carbon Project.
Grimston, M.C., Karakoussis, V., Fouquet, R., Vorst, R.V.D., Pearson, P. and Leach, M. (2001). The European and global potential of carbon dioxide sequestration in tackling climate change. *Climate Policy*, **1(2):** 155–171.
Hanagata, N., Takeuchi, T., Fukuju, Y., Barnes, D.J. and Karube, I. (1992). Tolerance of microalgae to high CO_2 and high temperature. *Phytochemistry*, **31(10):** 3345–3348.
Herzog, H.J. and Drake, E.M. (1993). *Greenhouse Gas R and D Program*. IEA/93/OE6.
Hsieh, H. (1990). Inorganic membranes. *Membr Mater Proc*, 84–91.
Idem, R., Wilson, M., Tontiwachwuthikul, P., Chakma, A., Veawab, A. and Aroonwiles, (2006). Pilot plant studies of the CO_2 capture performance of aqueous MEA and mixed MEA/MDEA solvents at the University of Regina CO_2 capture technology development plant, *Industrial and Engineering Chemical Research*, **45(8):** 2414–2420.
IEA (2016). "Hydropower - Fuels & Technologies", IEA, Paris
IEA (2022). Global CO_2 emissions rebounded to their highest level in history in 2021 - news. IEA.
IAEA (2016). Nuclear power and climate change.
Izadbakhsh, M., Gandomkar, M., Rezvani, A., Ahmadi, A. (2015). Short-term resource scheduling of a renewable energy based micro grid. *Renewable Energy*, **75:** 598–606. https://doi.org/10.1016/j.renene.2014.10.043.
Johnson, J.E. and Homme, A.C. (1984). Selexol solvent process reduces lean, high-CO_2 natural gas treating costs. *Energy Progress,* **4 (4):** 241–248
Kishimoto, S., Hirata, T., Iijima, M., Ohishi, T., Higaki, K., Mitchell, R. (2009). Current status of MHI's CO2 recovery technology and optimization of CO2 recovery plant with a PC fired power plant. *Energy Procedia*, **1 (1):** 1091–1098. https://doi.org/10.1016/j.egypro.2009.01.144.
Kovvali, A.S. and Sirkar, K.K. (2002). Carbon dioxide separation with novel solvents as liquid membranes. *Industrial Engineering Chemical Research*, **41:** 2287–2295.
Kozak, F., Petig, A., Morris, Ed., Rhudy, E., Thimsen, D. (2009). Chilled ammonia process for CO2 capture. *Energy Procedia*, **1 (1):** 1419–1426. https://doi.org/10.1016/j.egypro.2009.01.186.
Liu, Y.X., Jessop, P.G., Cunningham, M., Eckert, C.A. and Liotta, C.L. (2006). *Switchable Surfactants. Science,* **313(5789):** 958–960.

Majchrowicz, M.E., Brilman, D.W.F. and Groeneveld, M.J. (2009). Precipitation regime for selected amino acid salts for CO_2 capture from flue gases. *Energy Procedia 1*, 979–984.

Moorthy, K., Patwa, N. and Gupta, Y. (2019). Breaking barriers in deployment of renewable energy. *Heliyon*, **5(1):** e01166.

Nasirov, S., Silva, C., Agostini, C.A. (2015). Investors' Perspectives on Barriers to the Deployment of Renewable Energy Sources in Chile. *Energies*, **8(5):** 3794–3814. https://doi.org/10.3390/en8053794.

Nesamalar, J.J.D., Venkatesh, P. and Charles Raja, S. (2017). The drive of renewable energy in Tamilnadu: Status, barriers and future prospect. *Renewable and Sustainable Energy Reviews*, **73:** 115–124. https://doi.org/10.1016/j.rser.2017.01.123.

Ohunakin, O.S., Adaramola, M.S., Oyewola, O.M., Fagbenle, R.O. (2014). Solar energy applications and development in Nigeria: Drivers and barriers. *Renewable and Sustainable Energy Reviews*, **32:** 294–301. https://doi.org/10.1016/j.rser.2014.01.014.

Olorunfemi, T.R., Nwulu, N.I. and Ismail, S.O. (2022). Composites as candidate materials for photovoltaic cells. *Advances in Materials and Processing Technologies*, 1–20.

Painuly, J.P. (2001). Barriers to renewable energy penetration; a framework for analysis. *Renewable Energy*, **24(1)**: 73–89. https://doi.org/10.1016/S0960-1481(00)00186-5.

Qader, A., Hooper, B., Innocenzi, T., Stevens, G., Kentish, S., Scholes, C., Mumford, K., Smith, K., Webley, P.A., Zhang, J. (2011). Novel post-combustion capture technologies on a lignite fired power plant-results of the CO_2CRC/H_3 capture project. *Energy Procedia*, **4:** 1668–1675. https://doi.org/10.1016/j.egypro.2011.02.039.

Reddy, S., Scherffius, J.R., Yonkoski, J., Radgen, P., Rode, H. (2013). Initial Results from Fluor's CO_2 Capture Demonstration Plant Using Econamine FG PlusSM Technology at E.ON Kraftwerke's Wilhelmshaven Power Plant. *Energy Procedia*, **37:** 6216–6225. https://doi.org/10.1016/j.egypro.2013.06.550.

Renzenbrink, W., Ewers, J., Keller, D., Wolf, K.J., Apel, W. (2009). RWE´s 450 MW IGCC/CCS Project—Status and Outlook. *Energy Procedia*, **1:** 615–622.

Scholes, C.A., Kentish, S.E. and Stevens, G.W. (2009). The effect of condensable minor components on the gas separation performance of polymeric membranes for carbon dioxide capture. *Energy Procedia* **1:** 311–317.

Sen, R., Bhattacharyya, S.C. (2014). Off-grid electricity generation with renewable energy technologies in India: An application of HOMER. *Renewable Energy*, **62:** 388–398. https://doi.org/10.1016/j.renene.2013.07.028.

Sjostrom, S. and Krutka, H. (2010). Evaluation of solid sorbents as retrofit technology for CO_2 capture. *Fuel.* **89:** 1298–1306.

Song, C., Liu, Q., Deng, S., Li, H. and Kitamura, Y. (2019). Cryogenic-based CO_2 capture technologies: State-of-the-art developments and current challenges. *Renewable Sustainability Energy Review*, **101:** 265–278.

Sun, P., Nie, P-N. (2015). A comparative study of feed-in tariff and renewable portfolio standard policy in renewable energy industry. *Renewable Energy*, **74:** 255–262. https://doi.org/10.1016/j.renene.2014.08.027.

Suzuki, T.K. (2013) Evolution of solar-type stellar winds. *Astronomical Notes*, 334(1–2), Special Issue: Cool Stars, 17, 81–84. https://doi.org/10.1002/asna.201211751.

The Rogers (2011). Nuclear power plant accidents: listed and ranked since 1952.

Thiruvenkatachari, R., Su, S., An, H. and Yu, X.X. (2009). Post-combustion CO_2 capture by carbon fiber monolithic adsorbents. *Progress in Energy Combustion*, **35:** 438–455.

Timmerhaus, K.D. and Reed, R.P. (2007). Cryogenic Engineering: Fifty Years of Progress. Springer, Berlin/Heidelberg, Germany.

Tlili, T., Grevillot, G. and Vallieres, C. (2009). Carbon dioxide capture and recovery using TSA and/or VSA. *International Journal of Greenhouse Gas*, **3:** 519–527.

Van Hoist, J., Versteeg, G.F., Brilman, D.W.F. and Hogendoorn, J.A. (2009). Kinetic study of CO_2 with various amino acid salts in aqueous solution. *Chemical Engineering Science,* **64:** 59–68.

Wiatros-Motyka, M. (2016). An overview of HELE technology deployment in the coal power plant fleets of China, EU, Japan, and USA. *London: IEA Clean Coal Centre*, 75.

Wappel, D., Gronald, G., Kalb, R. and Draxler, J. (2010). Ionic liquids for post-combustion CO_2 absorption, *International Journal of Greenhouse Gas Control*, **4(3):** 486–494.

Zhang, H., Li, L., Zhou, D., Zhou, P. (2014). Political connections, government subsidies and firm financial performance: Evidence from renewable energy manufacturing in China. *Renewable Energy*, **63:** 330–336. https://doi.org/10.1016/j.renene.2013.09.029.

Zhang, J., Webley, P.A. and Xiao, P. (2008). Effect of process parameters on power requirements of vacuum swing adsorption technology for CO_2 capture from flue gas. *Energy Conversion and Management,* **49:** 346–356.

Zhao, C.W., Chen, X.P., Zhao, C.S. and Liu, Y.K. (2009). Carbonation, and hydration characteristics of dry potassium-based sorbents for CO_2 capture. *Energy Fuels,* **23:** 1766–1769.

Zhao Z, Cui X, Ma J, Li R (2007) Adsorption of carbon dioxide on alkali-modified zeolite 13X adsorbents. *International Journal of Greenhouse Gas Control,* **1:** 355–359.

Index

J. Marrugo-Negrete et al. (eds.), *Remediation and Carbon Capture in Mining Sector*, https://doi.org/10.1007/978-3-031-87883-1

MIX
Papier aus verantwortungsvollen Quellen
Paper from responsible sources
FSC® C105338

If you have any concerns about our products,
you can contact us on
ProductSafety@springernature.com

In case Publisher is established outside the EU,
the EU authorized representative is:
Springer Nature Customer Service Center GmbH
Europaplatz 3, 69115 Heidelberg, Germany

Printed by Libri Plureos GmbH
in Hamburg, Germany